The Institute of Biology's
Studies in Biology no. 140

Biology of Yeast

David R. Berry

Senior Lecturer in Applied Microbiology
University of Strathclyde

Edward Arnold

First published 1982
by Edward Arnold (Publishers) Limited
41 Bedford Square, London WC1 3DQ

British Library Cataloguing in Publication Data

Berry, David R.
Biology of yeasts.—(The Institute of Biology's studies in biology, ISSN 0537-9024; no. 140)
1. Yeast fungi
I. Title II. Series
589.2'33 QK617.5

ISBN 0-7131-2838-0

Photoset and printed by Photobooks (Bristol) Ltd

General Preface to the Series

Because it is no longer possible for one textbook to cover the whole field of biology while remaining sufficiently up to date, the Institute of Biology proposed this series so that teachers and students can learn about significant developments. The enthusiastic acceptance of 'Studies in Biology' shows that the books are providing authoritative views of biological topics.

The features of the series include the attention given to methods, the selected list of books for further reading and, wherever possible, suggestions for practical work.

Readers' comments will be welcomed by the Education Officer of the Institute.

1982

Institute of Biology
41 Queen's Gate
London SW7 5HU

Preface

In this book I have tried to introduce the reader to a wide range of studies which have been and are at present being carried out in one organism, Bakers Yeast. This single fungal species is used by every society in the world for the production of bread or alcoholic beverages, so it is a well recognized organism besides being of considerable economic importance. Yeast is also one of the most favoured organisms for the study of living systems. I have tried to indicate how the scientific study of yeast has arisen out of a need to control traditional fermentation processes such as brewing, and how the present scientific studies are leading to the development of new yeast-based industries. The reader might ask himself how much yeast or yeast products he has consumed to-day.

Although the book is primarily about yeast, the reader will gain some insight into recent techniques and concepts in the field of physiology, cell biology and genetics since the study of one organism like yeast draws on information from all organisms and has significance to all Biological Science.

Glasgow, 1982

D.R.B.

Contents

General Preface to the Series iii

Preface iii

1 Introduction: the History of Yeast 1
1.1 Early exploitation of yeast by man 1.2 Early scientific study of yeast 1.3 Development of yeast classification 1.4 Brewers and bakers yeast

2 The Architecture of the Yeast Cell 4
2.1 Cell morphology 2.2 The cell wall 2.3 The Cell membrane 2.4 The nucleus 2.5 Mitochondria 2.6 Other cytoplasmic structures

3 Nutrition and Metabolism of Yeast 10
3.1 Nutritional requirements 3.2 Growth of yeast 3.3 Metabolism of yeast 3.4 Regulation of metabolism

4 Cell Biology of Yeast 17
4.1 Chromosome structure 4.2 DNA replication 4.3 RNA 4.4 Protein biosynthesis 4.5 Regulation of enzyme production 4.6 Mitochondria

5 Cell Growth and Division 23
5.1 The nature of the cell cycle 5.2 Nuclear behaviour during the cell cycle 5.3 Cell wall synthesis 5.4 Bud formation 5.5 Cell synchrony techniques 5.6 Biochemical events of the cell cycle 5.7Genetics of the cell cycle 5.8 Control of the cell cycle

6 Sexual Reproduction 32
6.1 Plasmogamy 6.2 Sporulation 6.3 Sporulation, a model for sexual differentiation in eukaryotic cells 6.4 Spore germination

7 Genetics and Genetic Manipulation 41
7.1 Selection of genetic markers 7.2 Recombination 7.3 Genetics of mitochondria 7.4 Spheroplast fusion 7.5 Transformation and gene transfer 7.6 Virus-like particles in yeast

8 Yeasts in Industry 48
8.1 Role of yeast in the production of alcoholic beverages 8.2 Types of organoleptic compounds 8.3 Fermentation processes 8.4 Gasohol production 8.5 Bakers yeast and biomass production 8.6 Yeast derived products 8.7 Coda

Appendix 58

Further Reading 59

Index 60

1 Introduction: the History of Yeast

1.1 Early exploitation of yeast by man

Although man must have been subject to the ravages of disease-causing organisms throughout his evolution, yeast appears to be the first micro-organism to be used by man. Yeast could with justification be viewed as one of the many tools developed by early man. The first records of the use of yeast by man are concerned with the production of a type of acid beer called 'boozah' in 6000 B.C. Egypt. This beer was produced by the fermentation of a dough which was prepared by crushing and kneading germinated barley. The processes for producing beer and wine, and for the production of leaven bread probably developed in parallel over the next few thousand years. In 1200 B.C. Egypt the distinction between leaven and unleaven bread was well established and the use of a portion of yesterday's dough to inoculate today's bread or to inoculate wine fermentations was well established. From Egypt, the technology of brewing and baking passed to Greece and hence to Rome and the Roman Empire.

There is a shortage of records on brewing in the period following the fall of the Roman Empire. It is clear however that in the thirteenth and fourteenth centuries, brewing was well established in the monasteries of northern Europe. It has been reported that between 400 and 500 monasteries in Germany were active in producing beer during this period. As early as 1188, Henry II had levied the first recorded tax on beer in Britain.

The origin of distilled beverages leaves much to conjecture. There are reports of distilled beverages in China in 1000 B.C. and it is clear that whiskey distillation was well established in Ireland in the twelfth century. It is believed that the process of distillation probably came to Europe from the Middle East and this view is supported by the fact that the word alcohol is derived from arabic. Again the production of distilled beverages appears to have been associated with religious establishments and one of the earliest references to whisky in Scotland refers to production by a Friar John Cor in 1494 (Table 1).

1.2 Early scientific study of yeast

A knowledge of the yeast structure was dependent upon the discovery of the microscope and in fact the first description of yeast is attributed to Antonie van Leeuwenhoek in 1680. At this time however, there was no suggestion that the structure described as yeast was a living organism. It is difficult to establish who were the first scientists to suggest that yeasts were living organisms which caused the alcoholic fermentation observed in wines and beer. Vitalistic theories of

Table 1 Some key steps in the technological utilization and scientific study of brewers yeast. (Abstracted from *Yeast Technology* by G. Reed and H. J. Peppler, A.V.I. Publishing Company, 1973, and other sources.)

6000 B.C.	Evidence of brewing in Egypt
1000 B.C.	Consumption of potable distilled spirits in China
1192 A.D.	Whiskey production in Ireland
1200–1300	Breweries established throughout northern Europe
1680	Observation of yeast by Antonie van Leeuwenhoek
1832	Yeast recognized as fungi by Persoon and Fries
1838	Brewing yeast named *Saccharomyces cerevisiae* by Meyer
1839	Yeast spores described by Schwann
1863	Role of yeast in fermentations established by Pasteur
1866	Life cycle of yeast demonstrated by de Barry
1881	Pure cultures obtained by Hansen
1896	Scientific system for classifying yeast published by Hansen
1897	Fermentation by cell free extracts of yeast reported by Buchner
1934	Alternation of haploid and diploid phases in life cycle of yeast demonstrated by Winge
1943	Heterothallism in *Saccharomyes* reported by Lindegren

fermentation were proposed in the late eighteenth century and in 1818 Erxleben suggested that yeast was responsible for alcoholic fermentations. However, it is generally agreed that it was the work of Pasteur published in his *Etudes sur Vin*, in 1866 which established beyond doubt the role of yeasts in the fermentation of sugars to alcohol. This work represents a milestone in the development of microbiology. Another important milestone was the establishment of pure yeast cultures from single cell isolates by Hansen in 1881. The use of pure cultures has been fundamental to the development of the taxonomy and physiology of yeast and other micro-organisms. In 1897, Buchner obtained a cell free extract by grinding yeast which was capable of fermenting sugars to alcohol, and by doing so established one of the foundation stones of modern biochemistry. Subsequent work in this area made a significant contribution to the elucidation of the Embden-Meyerhof-Parnas (EMP) pathway. Since this time, yeast has been a favoured organism for a wide range of physiological and biochemical studies. Of special interest to those interested in alcoholic beverages was the establishment by Ehrlich in 1906 of the relationship between amino acid metabolism and the production of fusel alcohols, a key group of organoleptic compounds produced by yeast.

Early developments in the field of microbial genetics also arose out of studies on yeast. The alternation of haploid and diploid phases in the life cycle of yeast was established by Winge (1935) who subsequently went on to demonstrate the Mendelian segregation of genes during sexual reproduction in yeast. These studies opened the way for extensive studies in yeast genetics which have made a major contribution to our understanding of the nature of the genetic material and the mechanism of inheritance in eukaryotic micro-organisms.

1.3 Development of yeast classification

Although the characteristic budding form of yeast has been recognized since the description by van Leeuwenhoek in 1680, a more precise description and identification of yeasts has always presented a problem. Since the vegetative forms of most yeasts do not have any distinctive morphological characteristics, they are not readily identifiable by direct observation. Initially, the name *Saccharomyces* was applied to all yeasts isolated from alcoholic beverages and three species were recognized by Meyen (1837) by their origin; *Saccharomyces vini* from wine, *S. cerevisiae* from beer and *S. pomorum* from cider. Yeast sexual spores were recognized by Schwann in 1837 but only in 1870 was the genus *Saccharomyces* restricted to those yeasts which produced spores.

1.4 Brewers and bakers yeast

The genus *Saccharomyces* contains some forty species, each of which produce spherical to ellipsoidal cells by budding, produce ascospores in asci and are capable of the efficient conversion of sugars to alcohol. The most important *Saccharomyces* species in brewing was isolated in pure culture, and described by Hansen as *Saccharomyces cerevisiae* var *ellipsoideus* (Hansen) Dekker. Strains classified as *S. cerevisiae* have been widely used for brewing, distilling, wine making, the production of bakers yeast and biomass production. However the species has been defined in different ways at different times. A recent monograph published by Barnett, Payne and Yarrow (1900) includes in *S. cerevisiae* strains which have been classified in eighteen different species in previous systems of classification. One of these, *S. carlsbergensis*, is of particular interest, since it is the 'bottom yeast' used for the fermentation of lager. It is recognized by its ability to metabolize the sugar melibiose. Although widely referred to a *S. carlsbergensis* in the brewing industry, it was reclassified as *S. uvarum* by Van de Walt in 1970 and has been more recently included in the species *S. cerevisiae* by Barnett, Payne and Yarrow.

2 The Architecture of the Yeast Cell

2.1 Cell morphology

The cells of *Saccharomyces cerevisiae* are round, ovoid or ellipsoidal in shape and vary from 2.5–10 μm in width and 4.5–21 μm in length. Unstained cells exhibit little detail under the light microscope and even when inclusions in the cytoplasm are recognizable, it is difficult to know whether they represent vacuoles, granules or nuclei. Although more information can be obtained by using specific stains, it is only since the advent of the electron microscope that a clear picture of the yeast cell has emerged. The characteristic features of a typical yeast cell are shown in Fig. 2–1. It can be seen that the cell is bounded by a thick cell wall. Inside this it is possible to recognize many of the features of a typical cell; a plasmalemma, a nucleus, mitochondria, endoplasmic reticulum, vacuoles, vesicles and granules. The structure and function of these different structures will be considered in the remainder of this chapter.

The distinguishing feature of a growing population of yeast cells is the presence of the buds which are produced when the cell divides. The daughter cell is initiated as a small bud which increases in size throughout most of the cell cycle, until it is the same size as the mother cell. Most growth in yeast occurs during bud formation, so the bud is more or less the same size as the mature cell

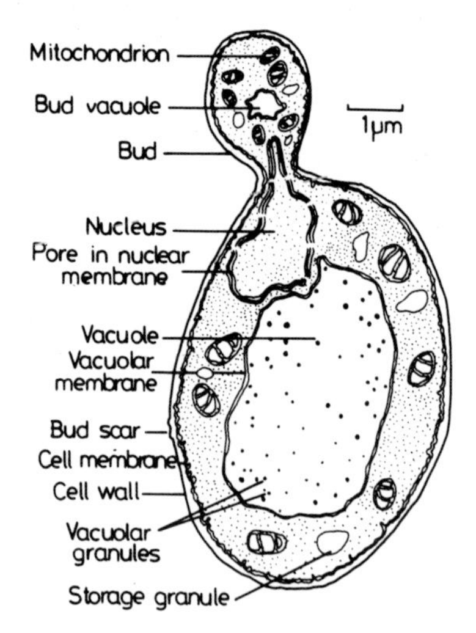

Fig. 2–1 Section through typical yeast cell showing the main features of the cell and their distribution. (Reproduced from Webster, 1980, *Introduction to Fungi*, p. 273. Cambridge University Press.)

before it separates. Cell separation may occur soon after cell division; however often new rounds of cell division take place before cell separation has occurred, so groups of cells are produced. The site of cell separation is marked on the mother cell by a structure referred to as the bud scar and on the daughter cell by the birth scar. These scars cannot be seen under the light microscope but can be seen using fluorescence microscopy after staining with fluorescent stains such as calcafluor or primulin. Bud scars and birth scars also show up as very distinct structures in scanning electron micrographs. No two buds arise at the same site on the cell wall in *Saccharomyces cerevisiae*, so each time a bud is produced a new bud scar is produced in the cell wall of the mother cell. By counting the number of bud scars, it is possible to establish the number of buds which have been produced by a particular cell. This can be used as a measure of the age of the cell. In any yeast population, 50% of the cells were produced by the last generation of cell divisions so possess a birth scar but no bud scar. Of the remaining 50%, 25% have one bud scar, 12.5% two bud scars and 12.5% more than two bud scars (Fig. 2–3c)

In some strains, cells growing in liquid culture adhere together to form clumps which settle to the bottom of the growth vessel. This phenomenon, which is referred to as flocculence, is of considerable importance in the brewing industry.

2.2 The cell wall

The cell wall is a rigid structure which is 25 nm thick and constitutes approximately 25% of the dry weight of the cell. Chemical analysis of the cell wall, indicates that the major components are glucan and mannan; however chitin and protein are also present. Glucan is a complex branched polymer of glucose units and is located in the inner layer of the yeast cell wall adjacent to the plasmalemma (see section 2.3). It appears to be the major structural component of the cell wall, since removal of the glucan results in a total disruption of the cell wall. Mannan, which is a complex polymer of mannose occurs mainly in the outer layers of the cell wall. Since it is possible to remove the mannan without altering the general shape of the cell, it appears that it is not essential to the integrity of the cell wall. The third carbohydrate component referred to; chitin, is a polymer of N,acetyl-glucosamine and is found in the cell wall of *S. cerevisiae* associated with the bud scars. Isolation of the bud scars by treating the cell wall with appropriate lytic enzymes, has shown that the chitin is arranged in a ring around the bud scar. Protein constitutes 10% of the dry weight of the cell wall. At least some of this protein is in the form of wall bound enzymes. Several enzymes have been described as being associated with the cell wall in yeast, including glucanase and mannanase, which are probably involved in the softening of the cell wall to permit bud formation; invertase, alkaline phosphatase and lipase. Several of these enzymes, e.g. invertase are mannoproteins and contain up to 50% of mannan, as an integral part of the enzyme molecule. Much of the remaining protein in the cell wall is also associated with mannan, so it is possible that this plays a structural rather than enzymic role in the cell wall. The detailed organization of the cell wall is not fully understood, but current theories favour a three-layered structure in which the inner glucan layer is separated from the outer mannan layer by a layer which is rich in protein (Fig. 2–2a).

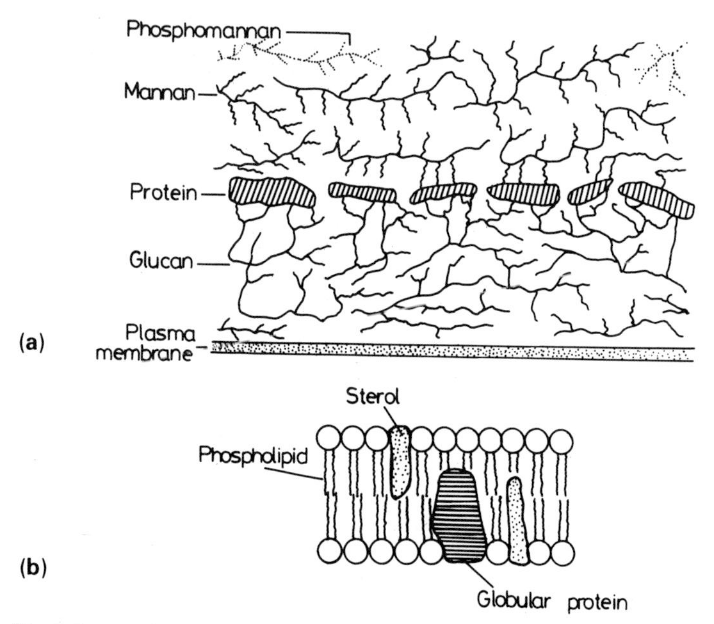

Fig. 2–2 (**a**) Diagrammatic representation of the structure of the yeast cell wall indicating the distribution of glucan and mannan in different layers. (**b**) Diagrammatic representation of a phospholipid membrane such as the plasmalemma. (Reproduced from Rose, A.H., 1976, *Chemical Microbiology*, p. 30. Butterworth, London.)

2.3 The cell membrane

The cell membrane or plasmalemma of the yeast cell can be observed using electron microscopy as a three-layered structure which is closely associated with the inner surface of the cell wall. It usually has a smooth appearance but at certain stages of the growth of the cell, invaginations can be seen. An understanding of the chemical composition of the plasmalemma requires isolation techniques which produce plasmalemma membrane free of other cellular components, including other membranes. One technique involves the formation of protoplasts; cells from which the cell wall has been removed by treatment with suitable lytic enzymes such as the snail juice enzyme, which is a preparation of lytic enzymes obtained from the gut of the snail *Helix pomatia*. Protoplasts remain intact if they are held in an isotonic solution of sugars but are readily burst when placed in more dilute suspension media. Different organelles, including the plasmalemma, can be obtained by centrifugation of a preparation of burst protoplasts. An alternative technique is to mechanically disrupt the cells first and remove the cell contents from the wall preparation by washing. The plasmalemma remains attached to the cell wall and can be liberated by dissolving the cell wall with lytic enzymes.

The plasmalemma is composed of lipids and proteins in more or less equal amounts, together with a small amount of carbohydrate. The main lipids present are mono-, di- and triglycerides, glycerophosphatides and sterols such as ergosterol and zymosterol. The nature of the protein in the plasmalemma is less well understood but probably includes the enzymes which are involved in the uptake of sugars and amino acids. Models of the structure of the plasmalemma have been presented (see Fig. 2–2b). Phospholipids are amphipathic molecules; that is each molecule consists of two regions, one of which is hydrophobic, i.e. repelled by water and another which is hydrophilic, i.e. attracted to water. They are believed to be arranged in such a manner that the hydrophilic parts of the molecule lie on the outside of the membrane and the hydrophobic lie on the inside of the membrane. The protein molecules may be arranged on the surface of the membrane or pass through the membrane.

The plasmalemma is a major organelle in the cell. It acts as a permeability barrier around the contents of the cell and controls the transport of solutes into and out of the cell. Strong evidence has also been presented that the plasmalemma is involved in the control of cell wall biosynthesis in yeast. *Saccharomyces cerevisiae* is unusual in that it cannot synthesize certain unsaturated fatty acids and sterols when it is grown in strictly anaerobic conditions, so these must be supplied in the medium if growth is to continue. Since the fatty acids and sterols supplied are incorporated into the cell membranes, it is possible to influence the chemical composition of the plasmalemma by feeding different fatty acids and sterols. Using this technique, it has been shown that changes in the lipid composition of the membrane affect the osmotic properties, the temperature sensitivity and the solute uptake characteristics of the cell.

2.4 The nucleus

The nucleus can be recognized by phase contrast microscopy and is usually situated between the vacuole and the bud. Chromatic bodies can be recognized in the nucleus using specific stains such as acid fuchsin or giemsa. However, knowledge of the studies of the yeast nucleus is limited, since individual chromosomes are very small, similar in size to a chromosome of *Escherichia coli* and not recognizable as discrete structures either by light microscopy or electron microscopy. The nuclear membrane remains intact throughout the cell cycle. It is visible in electron micrographs as a double membrane which is perforated at intervals with pores. Associated with the nuclear membrane is a structure referred to as a plaque which appears to function in a similar manner to the centrioles of animal cells. The characteristic structure of a plaque is a multilayered disc from which microtubules extend into both the nucleus and the cytoplasm. These plaques are considered to represent the spindle apparatus of the yeast nucleus and their behaviour has been monitored to follow the different stages of nuclear division (Chapter 4) (Fig. 2–3d).

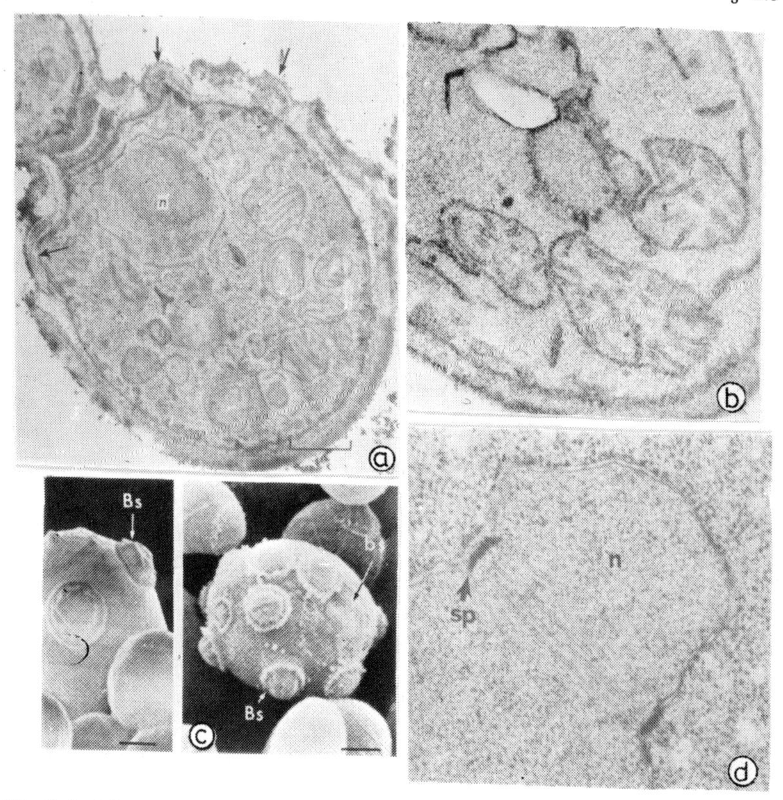

Fig. 2–3 Electron micrographs of yeast, showing sub-cellular structures. (**a**) Section of budding cell showing bud scars (arrows), nucleus (n), and developing cell septum. (Reproduced from Marchant and Smith, 1968, *J. Gen Micro.*, **53**, 168.) (**b**) Section of aerobic cell showing mitochondria. (Reproduced from Smith *et al.*, 1969, *J. Gen Micro.*, **56**, 54.) (**c**) Scanning electron micrograph showing bud scars (Bs) and birth scars (bs). (Reproduced from Belin, 1972, *Antonie van Leeuwenhoek*, **38**, 343.) (**d**) Section showing spindle plaques (sp) connected by microtubules during nuclear division. (Reproduced from Byers and Goetsch, 1973, *Cold Spring Harbour Symp.*, **38**, 123.)

2.5 Mitochondria

Mitochondria are readily recognizable in electron micrographs of aerobically grown yeast (see Chapter 3) as spherical or rod-shaped structures surrounded by a double membrane. They contain cristae which are formed by the folding of the inner membrane (Fig. 2–3b).

A considerable amount of work has been carried out on the structure of the mitochondrion and the distribution of the many mitochondrial enzymes in the membranes and the matrix of the mitochondrion. Most of the enzymes of the tricarboxylic acid cycle are present in the matrix of the mitochondrion, whereas the enzymes involved in electron transport and oxidative phosphorylation are associated with the inner membrane, including the cristae.

At one time it was considered that mitochondria were absent from anaerobically grown or catabolite repressed yeast (see Chapter 3) since they could not be detected and also because such cells lacked many of the enzymes associated with mitochondria. More recently, the use of freeze-etching techniques has indicated that the apparent absence of mitochondria was due to inadequate fixation techniques. Cells grown anaerobically in the absence of lipids have very simple mitochondria, consisting of an outer double membrane but lacking cristae. The addition of lipids such as oleic acid and ergosterol results in the development of cristae. The development of the mitochondrion is influenced by the lack of oxygen, the presence of lipids and the level of glucose in the medium. Thus, contrary to previous ideas, there is a change in the structure of mitochondria upon transfer from anaerobic to aerobic conditions but no *de novo* generation of mitochondria.

2.6 Other cytoplasmic structures

The cytoplasm of the yeast cell contains a system of double membranes known as the endoplasmic reticulum. Some of these membranes are associated with ribosomes as in other organisms; however, the endoplasmic reticulum appears to be involved in many other cellular activities. The relationship between endoplasmic reticulum and other organelles is not clear; however, there is continuity between the endoplasmic reticulum, the outer membrane of the mitochondrion and the plasmalemma. The endoplasmic reticulum is also involved in the formation of vesicles which are present in the cell. In this it behaves in a manner akin to the Golgi apparatus of some other organisms. It is not clear however, whether a true Golgi apparatus is present in yeast; membranous discs have been observed in yeast cells but they are few in number and not clearly recognizable as a Golgi apparatus.

Lipid granules are also present in the cytoplasm and again these appear to be derived from the endoplasmic reticulum.

Mature yeast cells contain a large vacuole; however at the point in the cell cycle when the bud formation is initiated, the vacuole appears to fragment into smaller vacuoles which become distributed between the mother cell and the bud. Later on in the cell cycle, these small vacuoles fuse again to produce a single vacuole in the mother and daughter cell.

The function of the vacuole is not well established. Evidence has been presented that it contains hydrolytic enzymes, polyphosphates, lipids and low molecular weight cellular intermediates, and metal ions. It may act as a storage reservoir for nutrients and for hydrolytic enzymes.

The technical problems of isolating and characterizing the different membrane components of yeast are considerable. Vesicles, vacuoles and other organelles are very fragile and easily disrupted. Since fragments of membrane from different organelles are almost impossible to separate, it is perhaps not surprizing that our understanding of the functional relationships between several of these structures is limited.

3 Nutrition and Metabolism of Yeast

3.1 Nutritional requirements

Although brewers and bakers yeast is probably the most readily recognized micro-organism for the layman, microbiologists have had difficulties in defining the species *Saccharomyces cerevisiae* in precise terms (Chapter 1). Since current opinion tends towards including several types of yeast which were previously classified as separate species, it is not easy to generalize on the nutritional requirements of *S. cerevisiae*. All the strains classified as *S. cerevisiae* can grow aerobically on glucose, sucrose, maltose and trehalose and fail to grow on lactose and cellobiose. However, growth on certain other sugars is variable, e.g. strains previously referred to as *S. carlsbergensis* or *S. uvarum* can metabolize raffinose completely and strains previously known as *S. diastaticus* and *S. chevalieri*, can grow on starch. The ability of yeasts to use different sugars can differ depending upon whether the cells are growing aerobically or anaerobically. Some *S. cerevisiae* strains cannot grow anaerobically on sucrose or trehalose.

All *S. cerevisiae* strains can utilize ammonia and urea as the sole nitrogen source but cannot utilize nitrate since they lack the ability to reduce it to ammonium ions. They can also utilize most amino acids, small peptides and bases as a nitrogen source. Histidine, glycine, cystine and lysine are however, not readily utilized. *S. cerevisiae* does not excrete proteases so extracellular protein cannot be metabolized.

The requirements of *S. cerevisiae* for growth factors such as pantothenate, biotin, thiamin, pyridoxine, niacin, folic acid and para-aminobenzoic acid are not clear, probably because different strains have different requirements, and also because the requirement for a growth factor can be influenced by other growth conditions. Yeasts also have a requirement for phosphorus, which is assimilated as the dihydrogen phosphate ion ($H_2PO_4^-$), and sulphur which can be assimilated either as sulphate (SO_4^{2-}) or as organic sulphur compounds such as methionine and cystine. The potassium requirement of yeast can be partly replaced by other alkali metals and even ammonium. Other metals required for good yeast growth include magnesium, calcium, zinc, iron and copper.

3.2 Growth of yeast

3.2.1 Batch culture

In many industrial processes such as brewing, distilling and wine making, yeast is grown in a medium which is rich in sugars (see Chapter 8). In these circumstances, growth is virtually anaerobic and the sugars assimilated are metabolized to carbon dioxide and ethanol by the process known as ethanol fermentation. Surprizingly, even if the vessel is well aerated, the sugars are still metabolized to carbon dioxide and ethanol until the sugar concentration has been reduced to a very low level. High levels of readily metabolizable sugars repress the ability of the cell to carry out aerobic respiration even when there is no shortage of oxygen. This phenomenon is known as *catabolite repression* and is very important in any understanding of the control of yeast metabolism.

Whereas in anaerobic conditions, the ethanol produced cannot be further metabolized, in aerobic conditions the yeast cells re-utilize the ethanol they have produced and oxidize it to carbon dioxide and water, once the sugars have been used up (see Fig. 3–1). It can be seen that in aerobic batch culture, growth stops when the glucose is depleted, then resumes after a short lag phase. During this lag phase, the cells are synthesizing the enzymes necessary for the aerobic metabolism of ethanol.

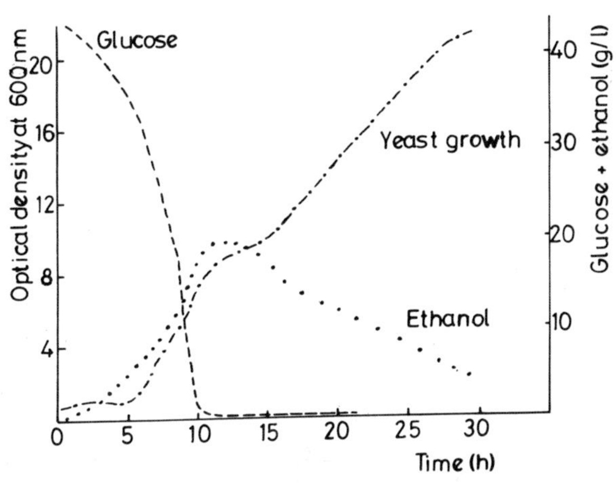

Fig. 3–1 Yeast growth in aerobic batch culture showing biphasic growth curve. An initial phase of fermentative growth on glucose is followed by a phase of oxidative growth on the ethanol produced in the fermentation phase. (Reproduced from Suomalainen, Nurminan and Oura, 1973, *Progress in Industrial Microbiology*, **12**, 150.)

S. cerevisiae can grow at a very rapid rate in fermentative conditions with a doubling time in the order of 1.6 h, however the yield of cells is very low. Similar growth rates can occur in conditions which permit aerobic metabolism and much higher yields are achieved. Since Pasteur was the first to note that a much higher yield of yeast cells could be obtained in aerobic conditions from a given quantity of sugar than in anaerobic conditions, this phenomenon is known as the *Pasteur effect*. This can be best illustrated by the growth of yeast in continuous culture.

3.2.2 Continuous culture

When cells are grown in batch culture, growth only continues for a short time before nutrients become depleted and growth stops. Continuous culture is a technique developed to permit cells to be grown at a constant growth rate for long periods of time. It is achieved by adding nutrients continuously to the growth vessel and removing spent medium together with the cells suspended in it, at the same rate. An equilibrium state is established, in which the number of cells produced by growth in the fermenter is balanced by the loss of cells in the washout from the fermenter. The growth rate can be controlled in such a fermentation by varying the rate of nutrient feed. If yeast is grown at different rates in continuous culture, the results shown in Fig. 3–2 are obtained. Below a growth rate of 0.2, all the glucose supplied is metabolized aerobically and no ethanol is produced. A high yield of yeast cells per unit glucose consumed is achieved in these conditions. Above a growth rate of 0.2, ethanol begins to accumulate and a decrease in yield of yeast cells is indicated by a decrease in the cell concentration in the fermenter. This occurs in conditions when the glucose concentration remains low. At even higher growth rates, glucose begins to accumulate since it is being supplied at a higher rate than it can be assimilated by the yeast cells. This experiment indicates that the yeast cell switches to fermentative metabolism at high growth rates even in conditions in which the glucose concentration remains low, and that this transition is associated with a loss of yield.

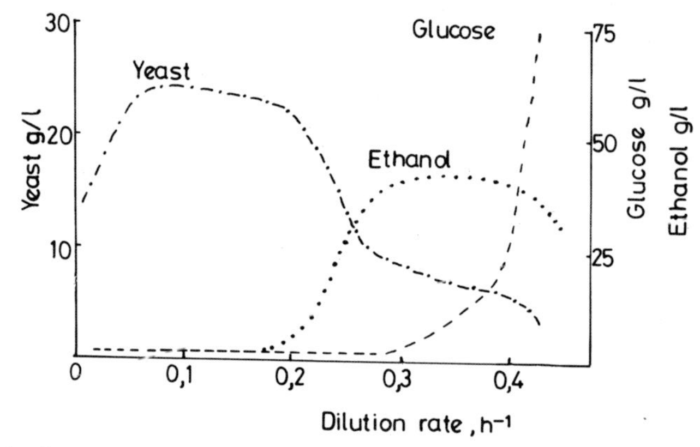

Fig. 3–2 Concentration of yeast, glucose and ethanol in continuous cultures of yeast grown at different dilution rates. The specific growth rate of yeast is approximately equal to the dilution rate. (Reproduced from Suomalainen, Nurminan and Oura, 1973, *Progress in Industrial Microbiology*, **12**, 151.)

3.3 Metabolism of yeast

3.3.1 Aerobic respiration

During aerobic growth on low levels of glucose, yeast metabolizes glucose to carbon dioxide and water in a manner similar to that which occurs in most plant and animal tissues (Figs. 3–3 and 3–4).

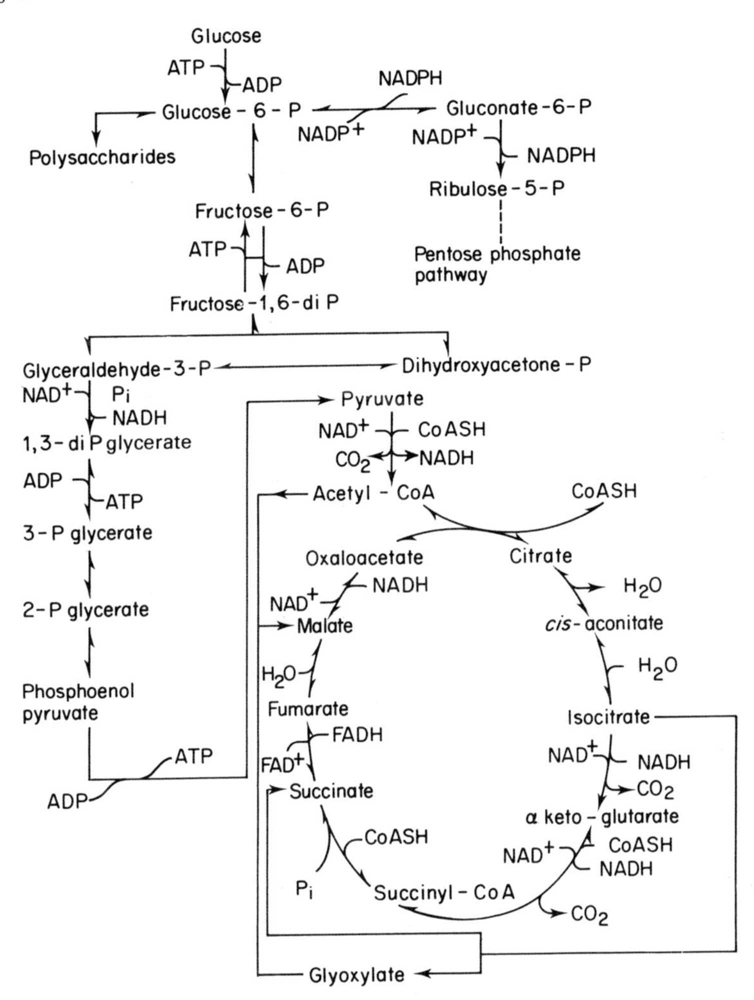

Fig. 3–3 Intermediary metabolism of yeast showing the glycolytic pathway, the pentose phosphate pathway and the tricarboxylic acid and glyoxylate cycles.

3.3.2 Fermentation

When yeast is grown in anaerobic conditions, glucose is metabolized to pyruvate, mostly via the EMP pathway. However, the pyruvate produced cannot be metabolized via the tricarboxylic acid (TCA) cycle and the $NADH_2$ generated during glycolysis cannot be oxidized via the cytochrome system. In these conditions, the pyruvate is decarboxylated by a different enzyme to produce acetaldehyde which is reduced to ethanol by the $NADH_2$ generated in glycolysis. In this way NAD is regenerated and can be re-utilized to permit the

$NADH_2 \rightarrow$ Flavoproteins $\rightarrow$ Ubiquinone $\rightarrow$ Cyt*b* $\rightarrow$ Cyt*c* $\rightarrow$ Cyta_3 ($\frac{1}{2}O_2 \rightarrow H_2O$)

NAD ↲

(a)

$$NADH_2 + \frac{1}{2}O_2 \rightarrow NAD + H_2O$$
$$3\ ADP + 3P_i \rightarrow 3\ ATP$$
$$NADH_2 + \frac{1}{2}O_2 + 3P_i + 3ADP \rightarrow NAD + 3\ ATP + H_2O$$

(b)

Fig. 3–4 (**a**) Intermediates involved in the transfer of electrons from reduced NADH to oxygen in yeast mitochondria. (**b**) Overall reaction involved in the generation of ATP during oxidative phosphorylation.

continuation of glycolysis. NAD can however be regenerated by other reactions such as the formation of glycerol from dihydroxyacetone phosphate (Fig. 3–5).

During the fermentation of glucose to ethanol, only four ATP molecules are generated, so fermentation is a much less efficient process than aerobic respiration. Since the amount of ATP required for biosynthesis is the same, whether the cells are growing aerobically or anaerobically, much more glucose is metabolized to produce the same quantity of cell material in anaerobic conditions. This provides a biochemical rationale for the Pasteur effect.

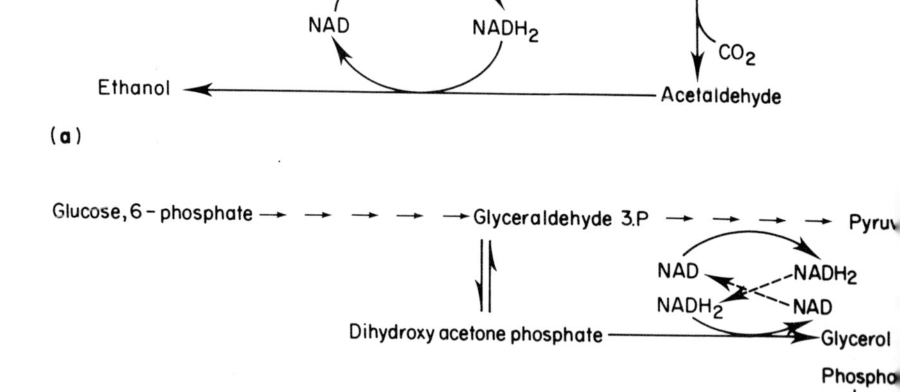

Fig. 3–5 Diagram of formation of (**a**) ethanol and (**b**) glycerol during fermentation indicating how NAD is recycled during these processes.

As previously indicated, alcohol fermentation does not only occur in anaerobic conditions. Yeast cells grown on high levels of glucose in aerobic conditions also ferment glucose to ethanol. The suppression of aerobic respiration by high levels of glucose or more correctly a high rate of assimilation of glucose is referred to as the Crabtree effect or catabolite repression. This suppression does not occur when *S. cerevisiae* is grown on less readily metabolizable sugars such as galactose. Catabolite repression of aerobic respiration does not only affect energy generation but also affects the supply of intermediates from the TCA cycle. In cells growing in a fermentative manner, both the TCA cycle and the glyoxylate cycle are suppressed. The TCA cycle acids required for biosynthesis are, however, produced by the carboxylation of pyruvate.

3.3.3 Gluconeogenesis

When all the glucose has been fermented to alcohol, the catabolite repression of aerobic respiration is removed so growth can continue by the aerobic metabolism of the ethanol produced in the fermentative phase. During these conditions, all the carbon is provided by ethanol and the formation of hexose sugars required for cell wall biosynthesis and other cell functions is achieved by a reversal of the EMP pathway. This process is known as gluconeogenesis.

3.3.4 Storage carbohydrates

Yeast produces two storage carbohydrates, glycogen and trehalose, which can account for up to 20% of the dry weight of the cell. Whereas glycogen is mostly produced in anaerobic conditions, trehalose can be a significant storage carbohydrate during aerobic growth.

3.4 Regulation of metabolism

The regulation of yeast metabolism is a very complex process. The cell may be growing aerobically or anaerobically, in a catabolite repressed or non-repressed state. Each of the different states of growth is characterized by different levels of the EMP pathway, the pentose phosphate pathway, the TCA cycle and the glyoxylate cycle. Elucidating the mechanisms which control the activities of these pathways in different growth conditions is a very difficult problem. Most of the steps in the glycolytic pathway involve readily reversible enzymic reactions; however, three steps, those catalysed by hexokinase, phosphofructokinase and pyruvate kinase are not readily reversible. These appear to play a critical role in the control of glycolysis. During gluconeogenesis, the reversal of glycolysis, these steps are catalysed by different enzymes or are bypassed by a different sequence of enzymes.

The rate of glycolysis can be regulated by the availability in the cell of substrates or cofactors such as NAD and ADP. However, the activity of several key enzymes can be influenced by the presence of other low molecular weight molecules present in the cell which act as enzyme activators or inhibitors. These regulatory compounds are not substrates, products or cofactors of the enzymes being regulated, but are often key intermediates of other steps of intermediary metabolism. It is not possible to go into a detailed discussion of this type of regulation; however its diversity is indicated by the type of cellular intermediates which are often involved in such regulatory processes. These include ATP, ADP, acetyl CoA, citrate, glucose 6-phosphate, 6-phosphogluconate, fructose diphosphate, phosphoenol pyruvate, aspartate and ammonia. Citrate and glucose, 6-phosphate appear to be important in the control of the Pasteur effect. Regulation can also occur as a result of changes in the quantity of a particular enzyme.

When yeast is grown anaerobically or in catabolite repressed conditions, the TCA cycle, the glyoxylate cycle and mitochondrial development are suppressed. Several mitochondrial enzymes are reduced in quantity, including the cytochromes and some of the TCA cycle enzymes. However, not all the enzymes of the TCA cycle are absent from the cell since the enzymes required to produce α-ketoglutarate from oxalacetate are present at a high enough level to produce the α-ketoglutarate required for glutamate biosynthesis.

The enzyme alcohol dehydrogenase offers an interesting example of control by catabolite repression. There are, in fact, three alcohol dehydrogenase enzymes in the cell, ADH, I, II and III. ADH I is concerned with ethanol production during fermentation and is not subject to catabolite repression. ADH II and III are both concerned with the oxidative metabolism of ethanol and are not produced when glucose is present in the medium. ADH III is a mitochondrial enzyme whereas the other two enzymes occur in the cytosol.

The mechanism of catabolite repression is not fully understood. Whereas oxygen seems to be important in controlling the level of cytochromes, the level of glucose appears to be critical in the control of the other enzymes. It is not believed that glucose itself acts as a repressor but that the level of some, as yet unidentified, product of glucose metabolism, is involved in the control of catabolite repression. In bacteria, convincing evidence that cyclic AMP is involved in the control of catabolite repression has been presented, and that this control operates at the level of gene transcription (see Chapter 4). The situation in yeast is not so clear. The level of cyclic AMP is depressed in catabolite repressed cells, however there is little direct evidence that cyclic AMP operates at the level of transcription and some results suggest that it acts at the level of translation or enzyme activation in yeast.

4 Cell Biology of Yeast

Although yeast is a unicellular organism it possesses the characteristics of a eukaryotic cell. This, combined with a well studied genetic system, makes it an ideal organism for studying the basic cell biology of the eukaryotic cell. Many people believe that *Saccharomyces cerevisiae* will replace *Escherichia coli* as the favoured organism for cell biological studies, since discoveries made in yeast are probably more easily related to the regulatory systems operating in man.

In saying that yeast is a eukaryotic organism, we are in fact making a very important statement about the structure of yeast chromosomes and about the mechanisms of RNA and protein biosynthesis which occur in yeast. One of the important points to be discussed in this chapter is the identification of these characteristics which indicate that yeast is a eukaryote.

It has already been shown (Chapter 2) that the yeast cell possesses a discrete nucleus which is bounded by a nuclear membrane, and that nuclear division appears to be achieved by a mitotic mechanism in which chromosomes divide and are separated at metaphase. Yeast also has a life cycle which involves plasmogamy and meiosis and an alternation of haploid and diploid phases (Chapter 6). These characteristics are all unique to eukaryotic organisms. Other characteristics such as the existence of mitochondria, the structure of the chromosomes and the nature of the protein synthesizing system will be discussed in this chapter.

4.1 Chromosome structure

The majority of the DNA in the cell resides in the chromosomes of the nucleus although DNA is also found in the mitochondria and in the form of small circles, referred to as 2 μm plasmids. Genetic studies over many years have indicated that the nuclear genetic material of yeast is located in several chromosomes rather than a single chromosome as has been found in bacteria. Data on gene linkage suggests that there are at least seventeen chromosomes in the yeast nucleus, however they cannot be readily identified using cytological techniques. This is perhaps not surprizing since the DNA content of the haploid yeast cell is only 1×10^{10} daltons which is $100\times$ lower than is present in man and only $10\times$ greater than that of *E. coli*. Since the DNA in yeast is divided between at least seventeen chromosomes then individual chromosomes are very small, many of them less than half the size of the chromosome of *E. coli*.

Estimates of the size of individual chromosomes have been made by extracting DNA from spheroplasts using a very delicate technique. The DNA is then centrifuged under controlled conditions in which the speed of sedimentation can be measured. Spheroplasts, which are cells from which the rigid cell wall has been removed, are used since it is much easier to isolate the DNA from them using gentle techniques which do not break the DNA. Since the purpose of these experiments is to determine the size of the natural chromosomal unit of DNA, it is essential that the DNA is not broken into two or more pieces during the extraction procedure. From the speed of sedimentation of chromosomes obtained in such experiments, values of between 1.1×10^9 and 1.9×10^9 daltons have been calculated.

Histones are basic proteins which have been found associated with the DNA of the chromosomes of eukaryotic organisms. They do not occur in bacterial chromosomes. Five classes of histones are generally recognized in the chromosomes of higher eukaryotes; H_1, H_{2a1}, H_{2a2}, H_{2b}, and H_3. In yeast three of these have been identified; however H_1 seems to be absent and there are doubts about the presence of histone H_3. The apparent absence of histone H_1 may be the result of its sensitivity to breakdown by proteolytic enzymes during extraction. However, histone H_1 is thought to be responsible for the condensation of chromosomes during nuclear division. Since such a condensation does not occur in yeast, its absence would not be surprizing.

The histones in eukaryotic chromosomes are not spread evenly along the DNA but are arranged in repeating units like beads on a thread. Each histone unit contains two molecules of H_{2a1}, H_{2a2}, H_{2b} and H_3. Segments of DNA are wrapped round the histone units and these are resistant to degradation by a DNAase enzyme whereas the DNA segments which pass between the histone units are readily degraded by this enzyme (Fig. 4–1). The presence of this characteristic structure has been demonstrated in yeast by treatment of isolated chromatin with this DNAase. Yeast chromatin appears under the electron microscope as broad threads which are 16–18.5 nm in diameter and have a 'knobby' appearance. This knobby appearance may be caused by the repeating histone units.

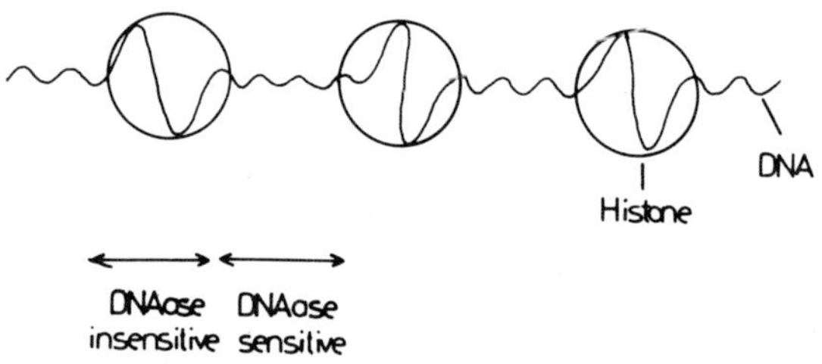

Fig. 4–1 Model of arrangement of spherical histone monomers along DNA of yeast chromosome. The DNA wrapped around the histone is more resistant to attack by DNAase than is the DNA between histone monomers.

4.2 DNA replication

The replication of nuclear DNA in yeast resembles that of other eukaryotes in that it occurs at several points along the chromosome at the same time. These replication points can be recognized as loops when chromosomal DNA is spread out and examined under the electron microscope. The junction between the replicated and non-replicated regions of the DNA form a characteristic Y form, the tail of the Y being the unreplicated double stranded 'parent' molecule and each arm of the Y consisting of one of the separated parental strands together with a newly synthesized daughter strand (Fig. 4–2). Such Y forms are visible in the majority of DNA molecules isolated from S phase cells (Chapter 5) but are absent from DNA isolated from G_1 phase cells in which the DNA is presumed not to be replicating. The existence of several sites of DNA replication which are situated approximately 30 μm apart on the chromosome has also been shown by a technique called autoradiography.

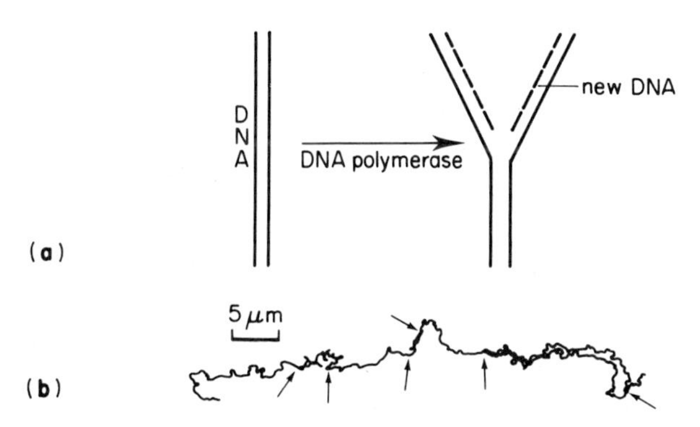

Fig. 4–2 (**a**) Diagrammatic representation of the replication of DNA showing how a forked structure can arise. (**b**) Tracing of an electron micrograph of replicating yeast DNA showing forks which arise as replication proceeds in both directions along the chromosome. (From Newlon, *et al.*, 1974, *Nature*, *London*, **247**, 32.)

4.3 RNA

As in other organisms, there are three types of ribonucleic acid (RNA) present in the yeast cell. The largest RNA molecules are the ribosomal RNA subunits which are incorporated into ribosomes. They are referred to as the large 23 S subunit and the small 16 S subunit. The S stands for Svedberg units which are a measure of the rate of sedimentation of molecules in a centrifuge. They give an indication of the size of the molecule. The ribosomal subunits of yeasts are larger than those found in bacteria and similar to those found in other eukaryotes.

Messenger RNA (mRNA) is a heterogeneous class of RNA molecules. It is responsible for carrying the coded information present in the DNA into the cytoplasm where it is used as a template in protein synthesis. Since there are thousands of proteins in the cell and a corresponding number of segments of DNA or genes in the nucleus, a similar number of unique mRNA molecules must be produced during the lifetime of the cell. Messenger RNA is produced in the nucleus and passes into the cytoplasm. However before this transfer occurs, a chain of adenosine molecules (poly A) is attached to one end of the mRNA molecule.

The third category of RNA which occurs in the yeast cell is called transfer RNA (tRNA). There are several different tRNA molecules in the cell, each of which has a specificity for one of the twenty amino acids which occur in proteins. They also have a specificity for one of the sequences of three bases in the mRNA which 'code' for a given amino acid. Because of this dual affinity both for a given amino acid and for the sequence of bases which codes for this amino acid in the mRNA the tRNA molecules are able to play a unique role as 'adaptor' molecules in translating the base sequence in the mRNA into a sequence of amino acids in the corresponding protein.

4.3.1 Biosynthesis of RNA

RNA is produced by a process known as transcription in which an RNA molecule is formed which has a base sequence 'complementary' to the sequence of bases present in the DNA from which it has been transcribed. The faithful matching of the sequence of bases is achieved by a system of base pairing which is similar to that which occurs in DNA replication. However in transcription it is complicated by the fact that thymidine present in DNA is replaced by uracil in the RNA molecule, so the pairing relationship in transcription is: A–U, G–C, T–A and C–G.

Three enzymes have been recognized in the yeast nucleus which are involved in RNA synthesis. They are referred to as RNA polymerase I, II, and III. The three polymerases appear to have different functions in the cell. RNA polymerase I is responsible for ribosomal RNA biosynthesis, RNA polymerase II for messenger RNA synthesis and RNA polymerase III for transfer RNA. In this respect also. yeast resembles other eukaryotic organisms.

After transcription has been completed, the structure of RNA molecules may be modified. The initial precursor of ribosomal RNA is cleaved into smaller subunits and modified by the methylation of certain bases before the final ribosomal subunits are produced. Many of the bases in transfer RNAs are also modified. Messenger RNA is also modified but in a different way. The poly A sequence referred to previously is added to one end of the molecule by the action of poly-A polymerase enzyme. Unlike other RNA segments, therefore, poly-A does not correspond to any sequence of DNA bases.

4.4 Protein biosynthesis

The mechanism of protein biosynthesis in yeast is essentially the same as that which occurs in other eukaryotes. The ribosome consists of a large 60 S subunit and a small 40 S subunit, rather than the much smaller subunits which occur in bacteria. Several ribosomes associate with a single mRNA to produce a structure which is described as a polyribosome. The mechanism by which the ribosomes attach to the mRNA is not fully understood in yeast; however once attached the ribosomes move along the mRNA translating the sequence of bases into a sequence of amino acids and producing a growing peptide chain as in other eukaryotes (see Fig. 4–3).

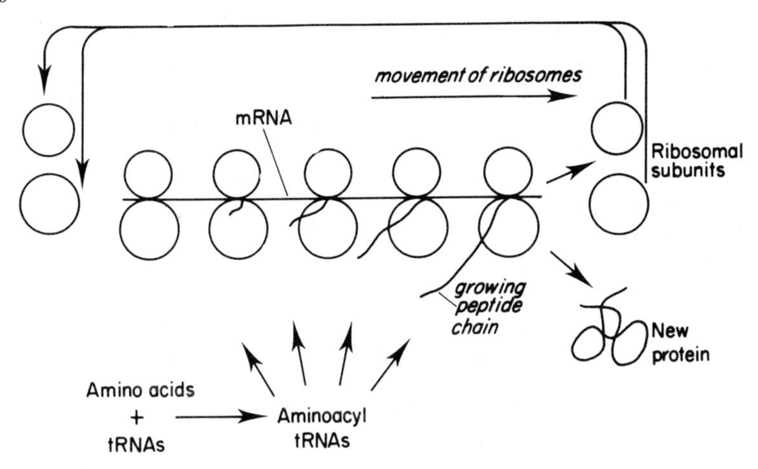

Fig. 4–3 Translation of messenger RNA into new protein.

4.5 Regulation of enzyme production

The level of enzymes present in the cell is influenced by many factors. Some enzymes are produced all the time and are referred to as constitutive enzymes. However, many enzymes are subject to some form of regulation which influences the level of the enzyme during different growth conditions or developmental states. The level of some enzymes increases in the presence of the substrate upon which they act or in the presence of a metabolically related compound. Galactokinase, galactose transferase and galactose epimerase are enzymes which are involved in the assimilation and metabolism of galactose. The level of these enzymes increases simultaneously when galactose is added to the growth medium. This regulatory process is known as enzyme induction. Other enzymes decrease in activity when the product of the metabolic pathway in which they occur is present in excess. Anthranilate synthetase is an enzyme which is involved in the formation of the amino acid, tryptophan. Its activity decreases when tryptophan is present in excess. This is known as enzyme repression. These two regulatory mechanisms, enzyme induction and repression operate on one or a few enzymes. However another regulatory mechanism which influences many enzymes operates in yeast. The levels of many enzymes, e.g. maltase and invertase, have a low activity in cells which have been grown on high levels of glucose, in excess of 1%. In these conditions the development of mitochondria and many of the enzymes associated with aerobic respiration are also repressed. This phenomenon is referred to as catabolite repression. It is evidently a very important regulatory process in yeast; however as yet the molecular mechanism by which it operates is not understood.

4.6 Mitochondria

Mitochondria contain their own DNA and protein synthesizing system and are therefore at least partly independent of the nucleus. The DNA in yeast mitochondria is a circular molecule which has a molecular weight of 50×10^6 daltons. This is five times as much DNA as is present in the mitochondria of higher animals. In being circular, the mitochondrial chromosomes resemble the chromosome of bacteria. It can be separated from the nuclear DNA because it is less dense and therefore bands at a different point on a density gradient. This is a centrifugation technique which separates molecules on the basis of density rather than size. It has been calculated that mitochondrial DNA constitutes between 15 and 23% of the DNA of the yeast cell which indicates that there are approximately fifty copies of the mitochondrial chromosome per cell. The mitochondrial DNA is replicated by a mitochondrial DNA polymerase which is distinct from the enzymes responsible for replicating nuclear DNA.

Ribosomes can be isolated from mitochondria which resemble bacterial ribosomes in size, having 39 S and 28 S subunits. These subunits contain 16 S and 12 S RNA species which have a base sequence complementary to segments of DNA in the mitochondrial chromosome and are therefore transcribed from this DNA. A unique set of tRNA molecules has also been found in yeast mitochondria. These can also be shown to be complementary to segments of the mitochondrial DNA. Thus the essential apparatus for protein synthesis is present in the mitochondria and is produced by them. Some mitochondrial proteins are also coded for by mitochondrial DNA. However, the majority of mitochondrial proteins are coded for by nuclear DNA and are synthesized on cytoplasmic as opposed to mitochondrial ribosomes. The site of biosynthesis of mitochondrial proteins can be distinguished by the use of inhibitors. One inhibitor, chloramphenicol, which is an inhibitor of bacterial protein synthesis, inhibits mitochondrial but not cytoplasmic protein synthesis in yeast. This illustrates the similarity between the mechanisms of mitochondrial and bacterial protein synthesis. A different inhibitor, cycloheximide, is active against cytoplasmic but not mitochondrial protein synthesis. By the use of these inhibitors it has been demonstrated that less than a quarter of the mitochondrial proteins are produced by the mitochondrial protein synthesizing system. There is a considerable amount of interesting work being carried out at the present time to determine which mitochondrial components are coded for by mitochondrial DNA and which by nuclear DNA.

5 Cell Growth and Division

5.1 The nature of the cell cycle

Growth in yeast is associated almost entirely with the growth of the bud which reaches the size of the mature cell by the time it separates from the parent cell. In rapidly growing yeast cultures, all the cells can be seen to have buds since bud formation occupies the whole cell cycle. In fact both mother and daughter cell can initiate bud formation before cell separation has occurred. In yeast cultures which are growing more slowly, cells lacking buds can be seen and bud formation only occupies part of the cell cycle. The cell cycle of yeast is normally

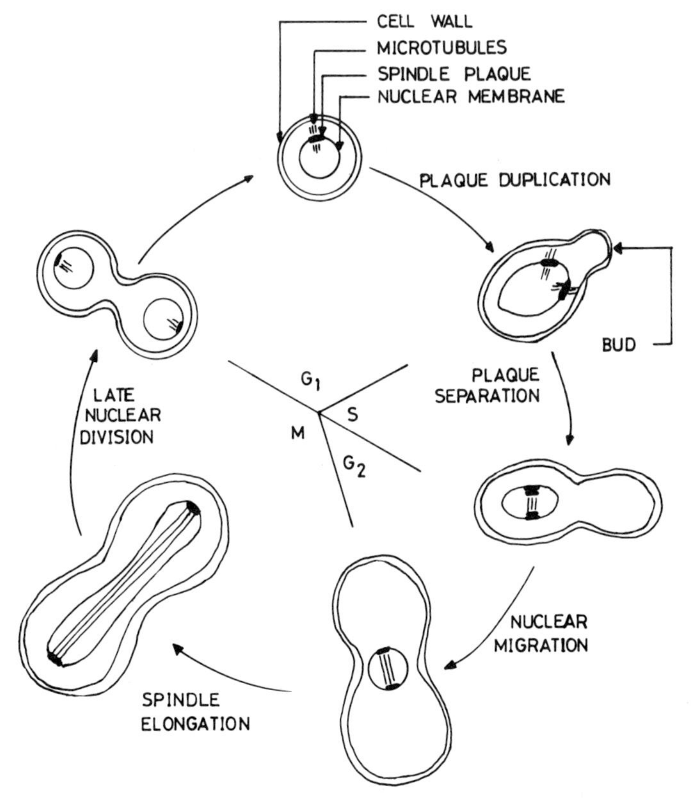

Fig. 5–1 Cytological events of the cell cycle indicating the relationship between bud formation, nuclear division and the phases of the cell cycle.

defined as the period between the end of one cell division and the next cell division. In cells which are growing in an unrestricted manner, all the contents of the cell double during this period. The sequence of events which occur during the cell cycle is shown in Fig. 5–1. The cycle is divided into four phases: G_1, S, G_2 and M. The S phase is the period when DNA synthesis occurs, the M phase is the time occupied by mitosis, and phases G_1 and G_2 represent the interval between mitosis and DNA synthesis (G_1) and DNA synthesis and mitosis (G_2). The onset of bud formation coincides with the initiation of DNA synthesis. The mechanism which controls the site of bud formation is not known: however it is possible that it is influenced by the orientation of the spindle plaques in the nucleus. The site of bud formation can nevertheless be identified before the bud appears by the appearance and increased number of sulphydryl groups in the cell wall. The completion of the cell wall between the bud and the mother cell, cytokinesis, occurs after the completion of mitosis; however cell separation may not occur at this time and may be delayed for several generations.

5.2 Nuclear behaviour during the cell cycle

During the S and G_1 phases of the cell cycle, the nucleus moves towards the site of the bud formation, so that at the onset of the M phase it is sited in the neck of the bud. Mitosis occurs in the neck of the bud in such a manner that when it is completed, one of the daughter nuclei has moved into the bud whereas the other remains in the mother cell.

As indicated in Chapter 1, it is not easy to recognize chromosomes in the nucleus of *S. cerevisiae* because the nuclear membrane remains intact during mitosis; however, by using electron microscopy it has been possible to identify different stages of the mitotic cycle by following the behaviour of the spindle plaques and the microtubules associated with them. These are considered to be components of the spindle apparatus which control chromosome movements during mitosis.

During the G_1 phase, a single plaque is visible in the nuclear membrane with short microtubules originating from it extending both into the nucleus and into the cytoplasm toward the point where the bud will develop. Towards the end of the G_1 phase, the plaque duplicates (stage PD) to produce two plaques which then separate (stage PS) and move apart until they are opposite each other on either side of the nucleus (Fig. 5–1). During this phase, microtubules are visible extending across the nucleus between the two plaques. The spindle remains in this state until the beginning of the M phase when the nucleus has moved to the neck of the bud. At this time the spindle elongates from 0.8 μm to 6.8 μm. The nucleus becomes stretched between the bud and the mother cell at this stage, before finally constricting and dividing into two nuclei. It seems probable that the dramatic separation of the two spindle plaques is important in separating two sets of chromosomes and moving one set into each cell.

5.3 Cell wall synthesis

The growth of the yeast cell wall occurs mainly during the growth of the bud; so we are dealing with the progressive increase in size of a rigid spherical structure. How this can be achieved is itself a fascinating problem which we have yet to solve. The yeast cell wall is also a very complex structure of which we have only a limited knowledge. Its biosynthesis must involve the formation of its several major components: glucan, mannan, chitin and protein, and their assembly into a three dimensional structure in a precise manner outside the cell membrane. The formation of the cell wall poses several interesting questions:

(1) What is the nature of the precursors from which the wall is synthesized?

(2) Which enzymes are involved in its biosynthesis?

(3) How do these enzymes control the three dimensional structure of the cell wall?

(4) Where does cell wall biosynthesis occur?

(5) At what stage in biosynthesis are cell wall components transported across the cell membrane?

The cell wall polysaccharides; glucan, mannan and chitin are produced from mannose, glucose and N-acetyl-glucosamine respectively. However, the immediate precursors of the polysaccharides are not the free sugars but uridine diphosphate (UDP) or guanosine diphosphate (GDP), derivatives of the sugars. The cell wall proteins are produced from amino acids by the normal process of protein biosynthesis (Chapter 4).

Very little information is available on the mechanism of glucan synthesis other than that cell free extracts can incorporate UDP glucose into glucan in the cell wall. However, several enzymes have been isolated which can add mannose units on to existing segments of mannan, using GDP mannose as the substrate. Genetic studies indicate that there are at least four different enzymes involved in the biosynthesis of mannan. It seems probable that the different kinds of bond found joining mannose units together in cell wall mannan (see Chapter 2) will be produced by different enzymes and it is possible to speculate that these enzymes control in some manner the nature and frequency of branching in the mannan. Enzymes which can transfer mannose from GDP mannose to mannan have been found to be membrane bound. At least some of these membranes appear to be derived from vesicles and it has been proposed that mannans may be synthesized inside vesicles and excreted through the cell membrane by a process which can be considered to be reverse pinocytosis.

One of the enzymes found in the cell wall, invertase, is a mannoprotein, that is it contains mannan as well as protein. Evidence has been presented that the protein component is synthesized in the cytoplasm by ribosomes of the endoplasmic reticulum and transported to the cell surface inside vesicles. At least part of the mannan present in the invertase molecule is attached while the enzyme is present in the vesicles.

There appear to be significant differences between the mechanism of glucan and mannan synthesis. Glucan synthesis can continue in the absence of protein synthesis and microfibrils of glucan can be seen on the surface of protoplasts which are being incubated in isotonic solutions of sugars. Mannan synthesis, on the other hand, cannot proceed in the absence of protein synthesis. Inhibitors of protein synthesis such as cycloheximide block mannan synthesis and mannan microfibrils do not accumulate on the surface of protoplasts. This dependence on protein synthesis has been interpreted as indicating that mannan synthesis can only be initiated by the attachment of mannose units to amino acids such as serine, threonine and asparagine in wall proteins.

5.4 Bud formation

The initial stages of bud formation involve the weakening of the cell wall caused by the action of lytic enzymes which attack the polysaccharides of the cell wall. The bud is formed by new cell material being laid down at the site of bud initiation then as bud formation progresses and the bud becomes larger, the deposition of new material becomes localized at the tip of the bud. When the bud reaches full size a complex septum is laid down in the neck of the bud which contains chitin in addition to glucan and mannan. Cell separation is achieved when the layers of the septum separate leaving the bud scar on the mother cell and the birth scar on the daughter cell.

The mechanism of bud formation has been subject to a series of elegant investigations by Cabib and his colleagues. They have studied the biosynthesis

CHITIN SYNTHETASE ZYMOGEN
ACTIVE CHITIN SYNTHETASE
ACTIVATING FACTOR
INHIBITOR

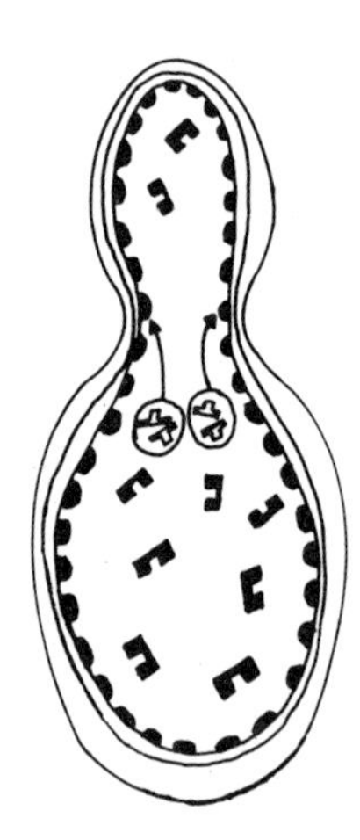
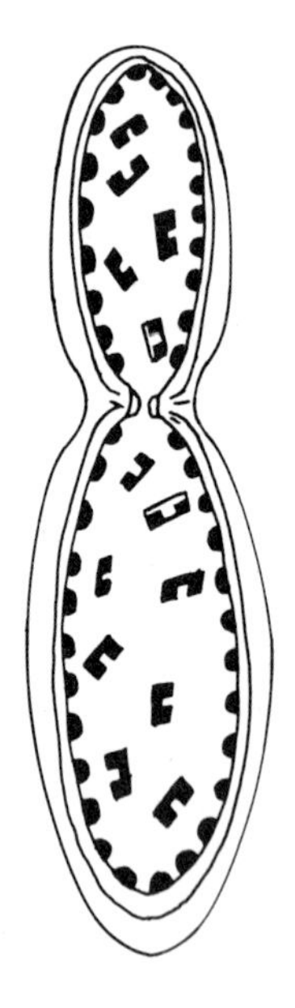

Fig. 5–2 Hypothetical scheme for the initiation of chitin synthesis. (Adapted from Cabib, E. and Farkas, V., 1971, *Proc. Nat. Acad Sci. U.S.*, **68**, 2052.)

of chitin which is found in the bud scar and represents one of the layers of the dividing wall formed during bud separation. The enzyme responsible for chitin synthesis, chitin synthetase, has been found to be present in the plasmalemma, as a zymogen, an inactive form of the enzyme. The formation of active chitin synthetase in the yeast cell is achieved by the action of an activation protein which has the properties of a proteolytic enzyme. Activation of chitin synthetase appears to involve proteolytic cleavage of the zymogen since it can be carried out *in vitro* using trypsin. The activation of chitin synthetase zymogen is restricted to the site of bud formation because the activator protein is not free in the cell but is contained in vesicles which appear to fuse with the plasmalemma in the region of bud formation. Thus the activation protein is only released at the site where it is required. The extent of the activation process is also restricted by the presence in the cytoplasm of an inhibitor of the activator protein which prevents it spreading throughout the cell and affecting other proteins (Fig. 5–2).

The development of many cellular structures requires that biosynthetic enzymes are active for a limited period of time at a restricted site in the cell. This work on chitin synthetase and bud formation in yeast provides one of the few models of how this can be controlled.

5.5 Cell synchrony techniques

Although some aspects of the cell cycle can be studied in normal cell populations containing cells at every stage of the cell cycle, other aspects can be studied more easily in synchronized cell populations in which the cells are at the same stage of the cell cycle at the same time. It is possible to achieve cell synchrony using a variety of techniques. These techniques can be divided into two types. In the first of these, cells are induced to grow synchronously by adding limited amounts of nutrient at regular intervals, The cells grow until the nutrients have been depleted then stop until a new addition of nutrient is made. Fortunately the cells always stop in the G_1 phase of the cell cycle so that when the cells start growing again they all start in this phase. If the process is repeated several times a high degree of synchrony can be achieved. This type of technique is known as *induction synchrony*. Similar results can be achieved by periodic alterations in the growth temperature or the pulse feeding of certain inhibitors of the cell cycle.

However, it is very difficult to interpret results obtained from cells synchronized in this manner. It is not possible to be certain whether any periodic change observed is the result of the cells being synchronous or the direct effect of the periodic treatment applied to obtain synchrony and therefore having no relevance to the natural cell cycle. Because of this problem, a second type of technique for obtaining synchronous cells is preferred. In this, the cells are grown up in a normal heterogenous population then centrifuged to separate them into different fractions based on size or density. Using this technique, populations of cells which are all at the same stage of the cell cycle can be obtained. If these are inoculated into fresh medium they grow synchronously for two to three generations. This technique is known as *selection synchrony*. When this technique is used, care must be taken that the centrifugation technique does not influence the results obtained.

5.6 Biochemical events of the cell cycle

A considerable amount of work has been carried out in *Saccharomyces cerevisiae* and other yeasts especially *Schizosaccharomyces pombe* in an attempt to decide when different cellular components are produced in the cell cycle. This may seem a relatively simple task but over the past decade many different and conflicting results have been obtained. One area of unanimity which exists is that DNA is synthesized during a limited period of the cell cycle, the S phase, producing a 'step' pattern of increase (Fig. 5–3). In contrast, the total protein and RNA increase continuously throughout the cell cycle. Many experiments have been carried out which indicated that many enzymes were produced in a step manner. The results of such experiments led to the view that different enzymes were produced at different times in the cell cycle and that a map of the cell cycle could be built up using the time of formation of different enzymes. It now appears that most step enzymes are an artefact, produced by the use of high osmotic strength solutions for gradient centrifugation during the selection synchrony procedure. If the selection is achieved using a continuous process of cell separation in a zonal rotor, then most enzymes are produced continuously throughout the cell cycle. The use of isotopic labelling techniques to study the time of formation of individual proteins also indicates the continuous formation of most cellular proteins. This work serves to illustrate how the results of an

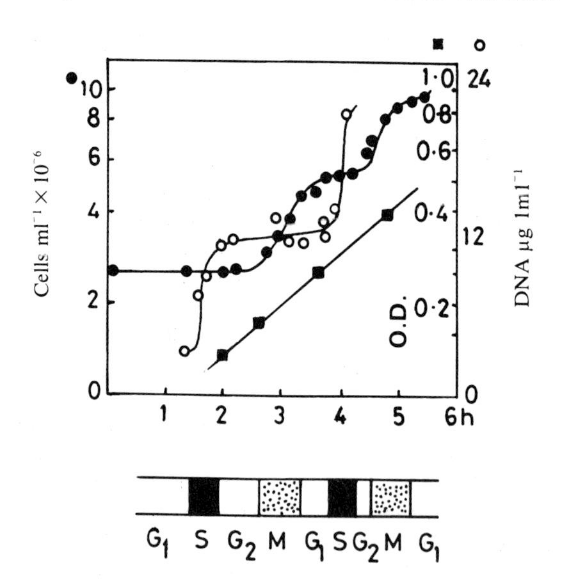

Fig. 5–3 Pattern of increase of protein, DNA and cell numbers during the cell cycle. The phases of the cell cycle are also shown.

experiment are only as reliable as the techniques employed and that even when all the obvious controls have been carried out, misleading results can be obtained as a result of factors which are not known to the investigator at that time. The formation of step enzymes under certain conditions remains a valid observation which still awaits explanation: however, they do not represent the normal pattern of enzyme synthesis during the cell cycle.

5.7 Genetics of the cell cycle

Our understanding of the cell cycle has been improved by the isolation of mutations which block different stages of the cell cycle. These are known as cell division cycle or *cdc* mutants. The isolation of mutations which block the cell cycle presents an interesting problem since cells blocked in the cell cycle cannot grow. This problem has been resolved by selecting temperature sensitive mutations which block the cell cycle at one temperature, the restrictive temperature, e.g. 37°C, but grow normally at another temperature, the permissive temperature, e.g. 25°C. When cells which have been growing at the permissive temperature are transferred to the restrictive temperature, they eventually stop growing. Such cells usually exhibit a characteristic morphology known as the terminal phenotype. However, this phenotype does not necessarily indicate the stage in the cell cycle at which the block had occurred, the execution point. This can be determined by careful control of the time of transfer of synchronous cells from the permissive temperature to the restrictive temperature.

A large number of *cdc* mutations have been isolated and studied. It has been found that they fall into two groups. The first of these groups contains mutations which affect functions such as DNA replication and nuclear division while the second group contains mutations which affect bud formation, growth and cell separation (Fig. 5–4). Whereas all the events affected by the mutations in the first group must occur before a new cell cycle can be initiated, the events affected by the mutations in the second group need not be completed before a new cell cycle is initiated.

5.8 Control of the cell cycle

When yeast cells are growing freely, the products of a cell division are two daughter cells which are identical to the original parent cell. It follows that all the individual components of the parent cell must have doubled during the period leading up to cell division; that is during one cell cycle. So far in this chapter, many of the individual events which occur in the cell cycle of yeast have been described, but little has been said as to how these events are co-ordinated so that cell growth and division proceed in an orderly manner and give rise to daughter cells which are similar in size, shape and biochemical characteristics to the mother cell.

Both biochemical and genetic studies suggest that the control of the cell cycle involves at least two different control sequences. One sequence is associated with DNA replication and mitosis and is referred to as the DNA division cycle and

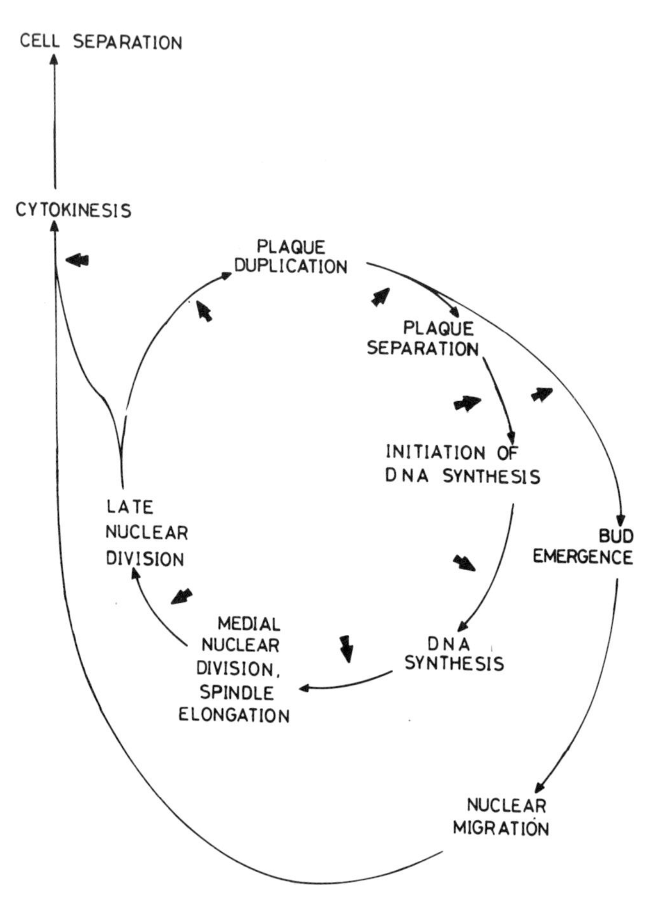

Fig. 5–4 Diagram showing site of action of mutations which affect the cell cycle in yeast (↑). The inner circle corresponds to the DNA division cycle and the outer sequence relates to the growth cycle. (From Hartwell, L.H., 1974, *Bacteriological Reviews*, **38**, 164.)

the other sequence, the growth cycle, is associated with the growth of the cell. Blocking the DNA division cycle with inhibitors or *cdc* mutations does not block the growth cycle and blocking the growth cycle with certain *cdc* mutations does not have an immediate effect on the DNA/division cycle. However, the observation that these two cycles can be separated does not mean that they do not normally function together. Although variations in cell size occur, the mean cell size of a yeast strain remains constant under most conditions. The important question to be answered is, 'How are growth and cell division linked'? Two situations can be considered.

(1) If the growth rate of a cell remains constant, the cell size would be controlled by the duration of the cell cycle. The more often cell division occurs the smaller the cells would be.

(2) If the duration of the cell cycle remains constant the size of the cell could be controlled by the rate of growth of the cell. The slower the rate of growth, the smaller the cells would be.

Evidence from both *S. cerevisiae* and *Schizosaccharomyces pombe* indicates that cell size is controlled by variations in the length of the cell cycle and not variations in growth rate. In *S. cerevisiae*, the variation in the cell cycle has been shown to be primarily due to variation in the length of the G_1 phase. These and other observations have been interpreted in terms of a model which proposes that the onset of cell division is initiated by an event in the G_1 phase called 'start' and that only cells which have achieved a given size can proceed beyond start. Such a mechanism would guarantee that only cells which had sufficient resources to complete the whole cell cycle would initiate the sequence leading to cell division.

6 Sexual Reproduction

Saccharomyces cerevisiae has a well characterized sexual cycle in which the haploid phase and diploid phase can exhibit vegetative development (Fig. 6–1). The haploid cell population consists of cells of one of two mating types, usually referred to as a and α although more recently the genes controlling these mating types have been named *mat* 1-A and *mat* 1-B respectively. Vegetative cells of opposite mating type can act as gametes and undergo plasmogamy and karyogamy to produce a diploid cell which may be grown vegetatively. However, if diploid cells are placed in an appropriate sporulation medium, haploid ascospores are produced by a process which involves a meiotic reduction in the number of chromosomes per nucleus. In this sexual cycle there are two phases which are unique to sexual reproduction, firstly plasmogamy and karyogamy, that is the fusion of *a* and α cells and the subsequent fusion of nuclei, and secondly ascospore formation. Each of these phases provides an excellent model for studying developmental processes in yeast and for studying the control of sexual reproduction in general.

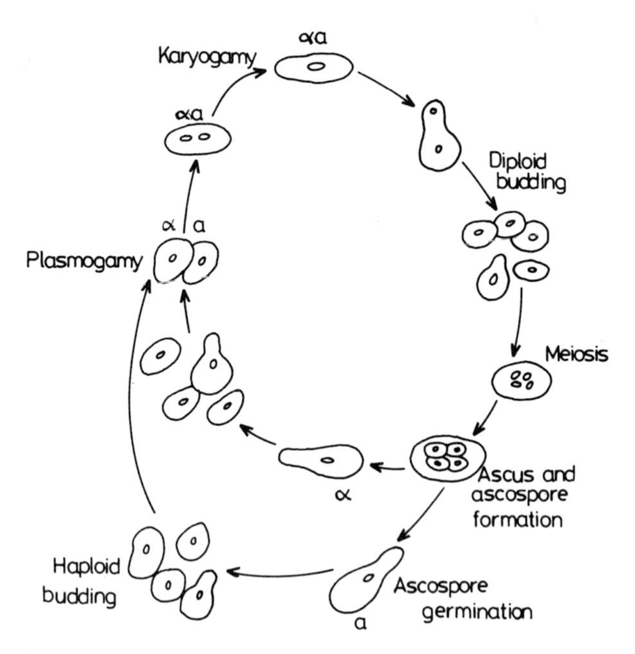

Fig. 6–1 Life cycle of yeast.

The importance of sexual reproduction is of course concerned with the recombination of the sets of genes from the two gametic nuclei. This aspect of sexual reproduction will be considered in the chapter on yeast genetics (Chapter 7). However, it should be noted that this process only has biological significance if the diploid strain is produced by the fusion of gametes from two genetically different strains. In yeast this is encouraged because plasmogamy can only occur between vegetative cells of opposite mating type, that is between *a* and α cells.

6.1 Plasmogamy

Plasmogamy occurs readily where *a* and α strains are mixed on a rich medium containing, for example, glucose, yeast extract and peptone (see Appendix 1).

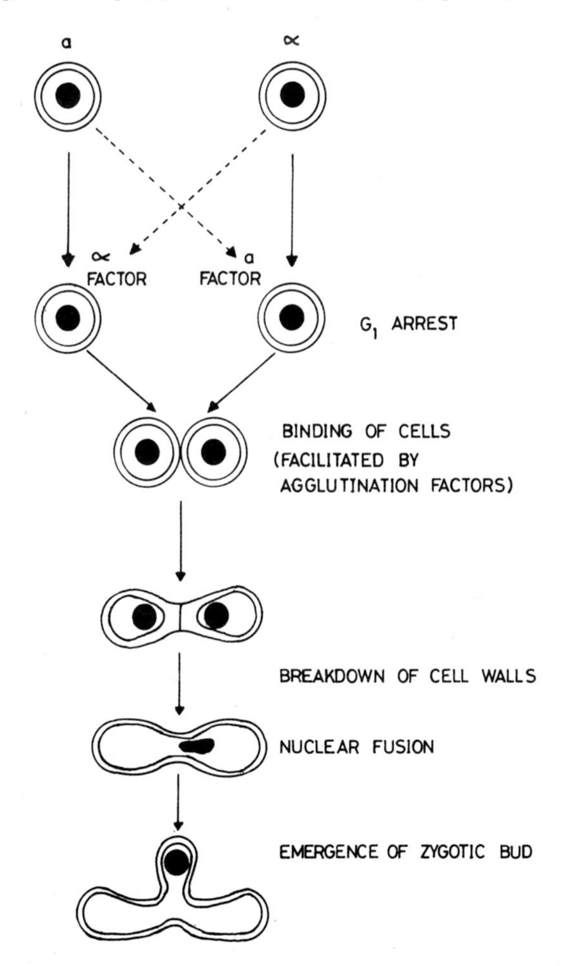

Fig. 6–2 Sequence of events during plasmogamy in *Saccharomyces cerevisiae*.

Within hours of mixing the two types of cell, characteristic dumbell shaped structures can be seen. However, not all the cells are able to mate initially, even if they are in contact since only cells at a certain stage of the cell cycle, the G_1 phase (see Chapter 5), are able to undergo plasmogamy, and only some of the cells will be in the G_1 phase. However, it is now known that *a* and α cells produce special peptides, α factor and *a* factor, which can bring the cell cycle of *a* cells and α cells respectively in the G_1 phase thus facilitating plasmogamy. These peptides are two of several pheromones produced by the *a* and α cells which are important in controlling the different stages of the complex process of plasmogamy.

Once contact has been established between an *a* and an α cell their cell walls become thinner in the region of contact. This appears to be mainly due to the breakdown of glucan in the cell wall in this area. As plasmogamy progresses, the cell wall and the cytoplasmic membranes from each cell become progressively disorganized such that cytoplasmic contact is established between the two cells. At this stage the cell can be considered as a transient heterokaryon, that is a single cell containing two genetically different nuclei (Fig. 6–2).

At this heterokaryotic stage the structure of the mitochondria from both mating cells becomes disorganized and it is believed that the processes which give rise to recombinant mitochondrial DNA occur at this point in the sexual cycle (see Chapter 7). However, this phase does not last very long; the two nuclei come into contact in the neck of the zygote and fuse to form the diploid nucleus. The movement of the nuclei and the site of nuclear fusion appear to be controlled by microtubules and spindle plaques similar to those observed in the vegetative cell cycle (Chapter 2). These same microtubules may also be responsible for controlling the site of the first zygotic bud, since this arises in the neck of the zygote close to the site of nuclear fusion. The development of the first zygotic bud is associated with a reorganization of the mitochondria and endoplasmic reticulum so that a normal vegetative cytoplasm is established. Budding then continues in a normal manner to establish a diploid population of cells.

6.2 Sporulation

6.2.1 Events of sporulation

Diploid cells of *Saccharomyces cerevisiae* are quite stable and, given suitable conditions, can continue growing and dividing indefinitely. However, if the diploid yeast is transferred from the growth medium to a sporulation medium containing, for example, just sodium acetate (see Appendix 1) cell division ceases and sporulation commences. During sporulation a special form of nuclear division occurs on which four haploid nuclei are produced. These four nuclei then become incorporated into four ascospores which develop inside the wall of the original diploid cell which has become modified to act as the ascus wall. The whole process of ascospore formation takes at least 15 h and may be prolonged up to 48 h depending upon the conditions used. Although sporulation is a much longer process than vegetative cell division, which can occur in less than 2 h, there are certain similarities between the two processes. During the initial period of sporulation the weight of the cell increases but the level of the DNA remains constant. After approximately 4 h, DNA replication occurs resulting in a doubling of the DNA level between 4 and 8 h. This is followed by a period in which nuclear division occurs. It must be emphasized again that this nuclear division is a meiotic dividion which gives rise to four haploid nuclei, not a simple mitotic division. The final stage of sporulation, the formation and maturation of the ascospores, can be compared to the cell division phase of the vegetative cell cycle, although again the mechanism is quite different. The similarity between the sequence of events in sporulation and those in the vegetative cell cycle is not the only evidence that the two processes are related. Both genetic and biochemical evidence has been presented that similar controls operate in the two processes (Fig. 6–3).

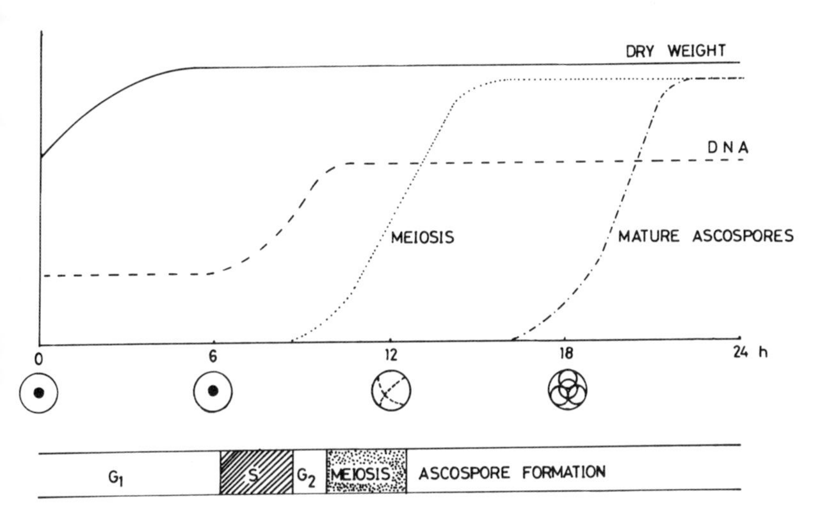

Fig. 6–3 Diagrammatic representation of sequence of physiological and cytological events of sporulation.

6.2.2 *Physiological control of sporulation*

Extensive studies have been carried out to identify the conditions which favour yeast sporulation. However, it is still not possible to describe the critical physiological parameters controlling the onset of sporulation and it is often difficult to obtain the high levels of sporulation which are desirable for both genetic and biochemical studies. The level of sporulation appears to be influenced by the conditions in which the cells have been grown as well as the conditions used for sporulation. The conditions required for sporulation are almost the opposite of those required for the pre-sporulation growth phase, The best results have been obtained using medium containing a high level of carbohydrates in the pre-sporulation phase together with a high level of a complex nitrogen source. It does not appear to be important whether the growth phase is aerobic or anaerobic. High levels of sporulation can also be obtained in yeast cells grown on a medium containing acetate as the carbon source. The highest levels of sporulation are normally obtained when non-growing cells are taken from the early stationary phase of the culture. These cells are in the G_1 phase of the cell cycle, and are also rich in glycogen. Glycogen is the storage compound produced by yeast which appears to provide a reserve of carbohydrate for sporulation.

In contrast, sporulation can only occur in aerobic conditions and on a very limited medium. This is normally a simple solution of sodium or potassium acetate. Sporulation is inhibited by the presence of readily assimilable carbohydrates such as glucose or sucrose and by nitrogenous material such as ammonia or amino acids. Addition of nitrogenous material leads to vegetative growth being re-established, whereas the addition of sugars simply represses sporulation. Growth is prevented by the lack of nitrogen. Sporulation can also be inhibited by the addition of ethanol.

It has been proposed that the natural conditions for sporulation in yeast involve the growth of the diploid yeast on the juice of fruit such as grapes. Since these are rich in sugars a fermentative mode of metabolism would be established resulting in ethanol production. When the carbohydrate has been fully utilized the yeast would metabolize the alcohol aerobically producing acetate. Once the level of ethanol has been reduced, suitable conditions for sporulation could arise.

During sporulation, acetate from the medium is assimilated and may be used as a carbon source. However, this may not be essential since the addition of acetate for a short period, 10 min, is effective in inducing sporulation. In these circumstances it appears to act as a trigger rather than acting as a carbon source. The lack of a requirement for nitrogen in the sporulation medium indicates that this not essential for sporulation and that any biosynthesis of nitrogenous compounds such as proteins and nucleic acids during sporulation must utilize stored nitrogenous compounds or recycled nitrogen from existing proteins and nucleic acids in the cell.

6.2.3 Cytology of sporulation

Although some changes occur in the cytoplasm of the ascus as it develops, for example, changes in the distribution of vacuoles and an increase in the frequency of oil droplets, the most interesting changes occur in the nucleus. The whole of the meiotic division occurs within the nuclear membrane. Since the nucleus remains more or less opaque in the light microscope throughout meiosis, progress in understanding the meiotic process in yeast has been dependent upon studies using the electron microscope. The thin sections used for electron microscopy do not give a clear view of chromosomes, so the events of meiosis have to be deduced from observations on the behaviour of the spindle plaques and the distribution of microtubules (see Chapter 3). The single plaque present at the early stages of sporulation divides as the two daughter plaques separate. At this point microtubules extend between the two plaques (Fig. 6–4). The orientation and separation of these plaques is considered to represent the metaphase and anaphase of the first nuclear division of meiosis. The plaques then divide again and the process is repeated; the two daughter plaques separating in a process which is considered to be equivalent to the metaphase and anaphase of the second division of meiosis. Towards the end of the second division, the nuclear membrane becomes lobed and one plaque can be observed in

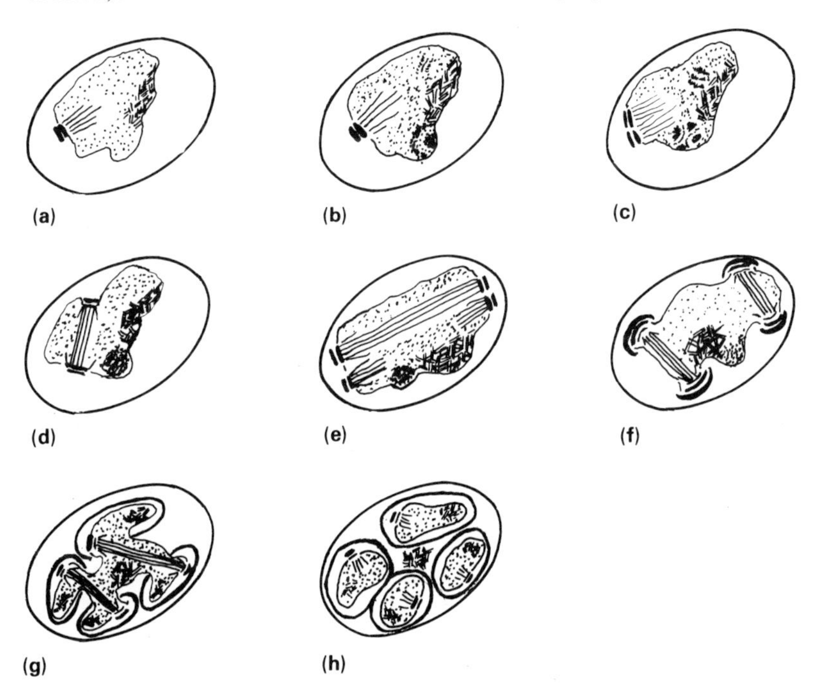

Fig. 6–4 Cytological events of nuclear division during meiosis and sporulation. The sequence of diagrams illustrates spindle elongation at first (a-d) and second (e-f) meiotic divisions and the formation of the ascospores (g-h). (Reproduced from Beckett *et al.*, 1974, *Atlas of Fungal Ultrastructure*. Longman, London.)

each lobe. The ascospore wall arises by the deposition of wall material around these lobes of the nucleus. The developing ascospore wall slowly encloses the whole of the nuclear lobes so that the original nucleus becomes divided into four nuclei which are enclosed in the developing ascospore wall. (Figs. 6–4 and 6–5b).

Ascospores in *S. cerevisiae* are spherical and have a thicker and stronger cell wall than vegetative cells. There are normally four ascopores per ascus arranged in a tetrahedral manner; however, two and three spored asci also occur and different arrangements can be observed (Fig. 6–5a).

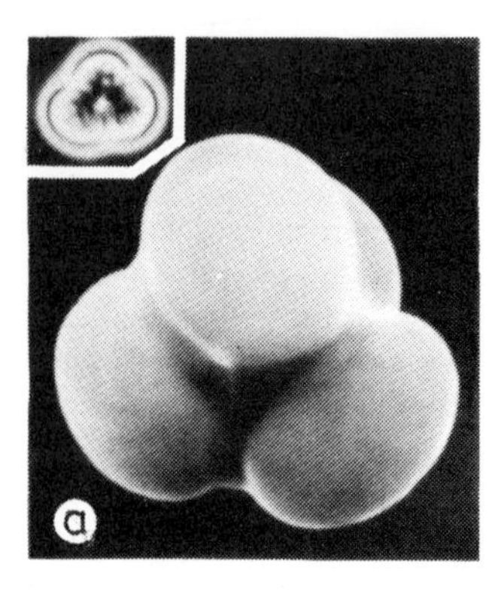

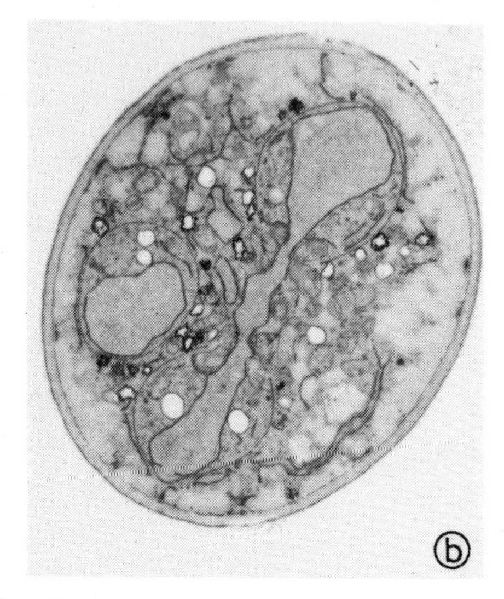

Fig. 6–5 Sporulating yeast cells. (**a**) Scanning electron micrograph of mature ascus. Inset is same stage photographed using phase contrast microscopy. (Reproduced from Rousseau *et al.*, 1972, *J. Bact*, **109**, 1235.) (**b**) Section through ascus showing development of ascospores and nuclear division. (Provided by Dr A. Beckett.)

The four ascospores normally remain in the ascus which only breaks down slowly in the natural environment. In the laboratory it can be broken down rapidly by treatment with glucuronidase enzymes, so releasing the ascospores. The ascospore wall is antigenically different from the vegetative cell wall and has a greater affinity for non-aqueous solvents. This property can be exploited to separate out ascospores from asci and vegetative cells for use in genetic studies.

6.2.4 Biochemistry of sporulation

It is apparent that the major cytological changes which occur during sporulation must be dependent upon considerable biosynthetic activity in the ascus. It is of considerable interest to know whether this biosynthetic activity involves the formation of enzymes which are specific to sporulation; however, as yet few individual proteins have been identified which are unique to the sporulation process. Although protein and RNA synthesis are dependent upon endogenous nitrogen reserves, the level of RNA and protein actually increases in the early stages of sporulation then decreases.

The energy required for biosynthesis appears to be derived from glycogen and glucan which decrease during sporulation. Trehalose on the other hand, another storage carbohydrate of yeast, increases during sporulation and is believed to act as the reserve carbohydrate in the ascospores which can be utilized during spore germination.

As has been indicated previously, the level of DNA doubles during ascus development. Since this doubling occurs during a restricted period of sporulation, the increase occurs in a stepwise manner rather than continuously. The biochemical mechanism of genetic recombination is not well understood; however, the available evidence indicates that recombination is associated with the period of DNA synthesis and is completed before the events of nuclear division occur. Synaptonemal complexes, structures which have in a variety of organisms been associated with recombination, have been reported in the yeast nucleus.

6.2.5 *Genetics of sporulation*

Since only diploids produced from a fusion of *a* and α cells can be induced to sporulate, it is not surprizing the genes controlling the *a* and α mating types are important in controlling sporulation. Using a variety of experimental techniques, it is possible to produce diploids which have two copies of the same mating type gene, i.e. they are *aa* or αα diploids. These strains, which do not sporulate, provide a useful tool for studying which changes are essential for sporulation and which are simply physiological changes in response to the transfer of cells from the growth medium to the sporulation medium. Using this technique, it has been shown that premeiotic DNA synthesis and the breakdown of glycogen only occur in *a*α diploids which are capable of sporulation and do not occur in *aa* and αα diploids. The mechanism by which this control of DNA biosynthesis is exerted is not known at present.

In the discussion of the vegetative cell cycle (Chapter 5), it was indicated that several mutants, called *cdc* mutants, have been isolated which block the cell cycle and nuclear division. Studies have been carried out to indicate whether the mutations which affect mitotic nuclear division, also affect meiotic nuclear division. The results obtained showed that all mutations which affected the vegetative cell cycle affected sporulation, except those which acted on bud formation, and cell separation. This is strong evidence that the control processes of the vegetative cell cycle and the meiotic cell cycle which occurs during sporulation have much in common.

Many other mutations have been isolated which block sporulation or give rise to abnormal asci; however, a great deal of work is still required before we have a clear idea which genes control sporulation in yeast, and how they operate.

6.3 Sporulation, a model for sexual differentiation in eukaryotic cells

Differentiation can be defined as a stable development of an altered structure and function. One of the ultimate challenges of biology is to understand how differentiation occurs. This challenge can be rephrased in a more mechanistic manner; how can cells which contain the same genetic information express it in different ways to give rise to different cellular types? In most fungi and in other organisms the process of sexual reproduction is associated with the formation of unique, differentiated structures for the production of gametes or meiotic spores. In many organisms these structures only constitute a small percentage of the mycelium and so are difficult to study. Since in yeast it is possible to induce sporulation in over 80% of the cell population, yeast offers an ideal model system for the study of this phase of the sexual cycle. Such studies are also facilitated by the wealth of biochemical and genetic information available in yeast.

Any differences which arise in cells of the same genotype must be the result of a change in environmental influences, a change in the way the genes are expressed, or interaction between the two. In yeast it can be shown that the onset of sporulation is dependent upon the correct physiological conditions being provided. Limited progress has been made in identifying which genes are involved and which biochemical changes occur during sporulation. The observation that several genes which affect the cell cycle also affect sporulation suggests a similarity of control mechanism between the two processes. However, whereas cells of all mating types, *a*, α, *aa*, αα, and *a*α, can undergo vegetative cell divisions, only diploid cells with both *a* and α, undergo sporulation. These genes appear to control the transition from vegetative growth to both plasmogamy and sporulation.

In any differentiating system there must be a starting point when the normal vegetative development of the cell starts to deviate to produce new structures. In yeast, the start signal for the cell to cease budding and initiate sporulation always occurs in the same phase of the cell cycle, the period between the last mitotic nuclear division and the subsequent period of DNA synthesis, the period known as the G_1 phase in the vegetative cell cycle. Sporulation can be reversed for several hours after this phase by the addition of nutrients to the sporulation medium.

It should be noted that plasmogamy in yeast also occurs in the G_1 phase and that yeast cells which stop growing always stop in the G_1 phase. The G_1 phase appears to be a very critical phase in the cell cycle from which the cell can develop in different directions depending upon the genetic composition of the cell and the nutrient status of the medium.

6.4 Spore germination

The germination of yeast ascospores has been less intensively studied than ascospore formation; however it is clear that the conditions which promote germination are different from those which favour growth or sporulation. Rapid germination only occurs in the presence of a readily assimilable carbon source such as glucose or fructose. It can occur in both aerobic and anaerobic conditions.

7 Genetics and Genetic Manipulation

Genetics has been defined as the study of inheritance. The developments in the field of yeast genetics and molecular biology have their origins in the work of Winge and his colleagues between 1935 and 1940 who established that the life cycle of yeast involved an alternation of haploid and diploid phases. This was followed by the demonstration by C.C. and G. Lindegren that mating in yeast was controlled by two alleles of a single gene referred to as *a* and α. Since this time, extensive and varied research has been carried out into the genetics of yeasts involving thousands of research workers so it is therefore only possible to outline the essential features of the genetic system of yeast in this chapter.

It is legitimate to include in genetics a variety of topics from the structure of the genetic material and the biochemical mechanisms involved in gene expression (Chapter 4) to the laws governing the inheritance of genetic information and the cellular mechanisms upon which these laws are based.

Genetics can also be used as an analytical tool to increase our understanding of all aspects of cellular function such as the control of enzyme formation, DNA replication, morphogenesis and the cell cycle. Since the ultimate control of all cellular activities lies with the genome, any description of cellular regulation which omits genetic control is incomplete. Some of these topics have been mentioned in other chapters. In this chapter the mechanisms controlling the inheritance of genetic information and the techniques used to study yeast genetics will be emphasized.

7.1 Selection of genetic markers

An essential prerequisite of all genetic studies is the identification of genetic markers. These take the form of characteristics which take one form in one strain and another form in a different strain and which are controlled by genetic rather than environmental factors. Such characteristics arise spontaneously in natural populations although their frequency is often low. Limiting studies to such naturally arising markers would be very slow and laborious. Fortunately there are well known techniques for increasing the frequency of changes in the genetic material. These changes which are known as mutations can be used as genetic markers. These techniques involve a two step procedure. The first step is the treatment of the yeast cell with a physical agent, e.g. ultraviolet light, or a chemical agent, e.g. ethyl methyl sulphonate which are known to cause changes in the genetic material, that is the cellular DNA. These agents are known as mutagens. Even when these techniques are used, the frequency of mutation in a particular gene is still low. usually 1% or less, so the second step involves the development of a selective technique to identify those cells in which a particular mutation has occurred.

7.2 Recombination

7.2.1 During sexual reproduction

The majority of new mutations which occur naturally are deleterious to existing strains so mutation is probably not the major source of variation in natural populations. A much higher level of variation is obtained if the genomes of two different strains of yeast can be recombined such that new combinations of existing genes are produced. This can be achieved in many strains of yeasts by the process of sexual reproduction. A description of the process of sexual reproduction in yeast and the incompatibility mechanisms which operate to encourage the fusion of strains with different genotypes has been presented in Chapter 6. The critical stage of the sexual cycle which provides a mechanism for producing new combinations of genes is the process of meiosis. The results obtained by genetic analysis indicate that the process of ascospore formation involves a nuclear division which is functionally identical to the meiotic division described in other organisms. The procedure for genetic analysis by sexual reproduction in yeast is essentially simple. If *a* and α strains containing different genetic markers are spread together on the surface of complete medium both will grow and some *a* and α cells will fuse to form diploid cells. The diploid cells can be isolated by micromanipulation or by using appropriate selective markers, allowed to grow and reproduce on another complete medium plate, then

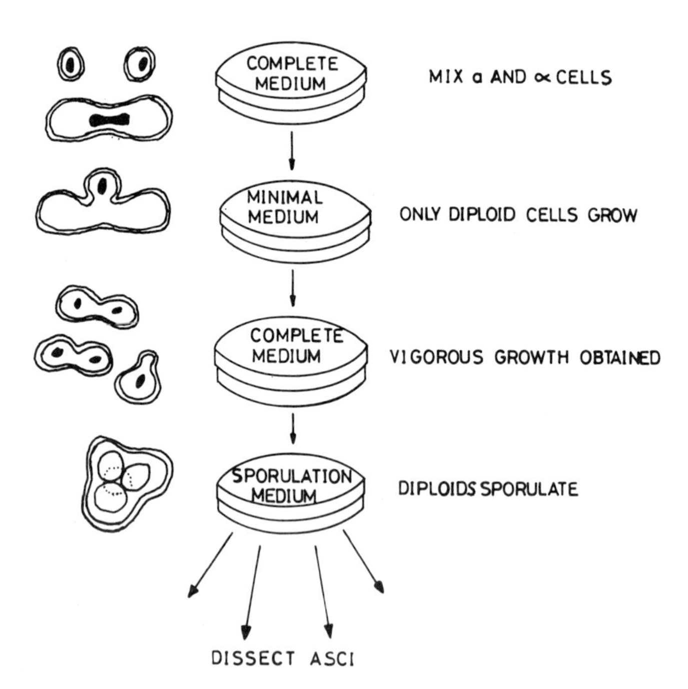

Fig. 7–1 Procedure for genetic studies using sexual reproduction in yeast.

transferred to sporulation medium. After 24–48 h some asci containing four ascospores should appear, although the percentage of diploid cells which produce four spored asci can vary from 80% down to zero. The four ascospores are the four products of meiosis so if they can be removed from the ascus and studied individually it is possible to see if any recombination has occurred between the different genetic markers used. The ascospores can be released by treating the ascus wall with an enzyme which was originally obtained from the gut of the snail *Helix pomatia* called 'snail juice enzyme', (Fig. 7–1). Other enzymes are now available, however snail enzyme is still widely used. Once the ascus wall has been dissolved, the ascospores, which are more resistant to these enzymes, can be dissected out and grown up as separate colonies which can be

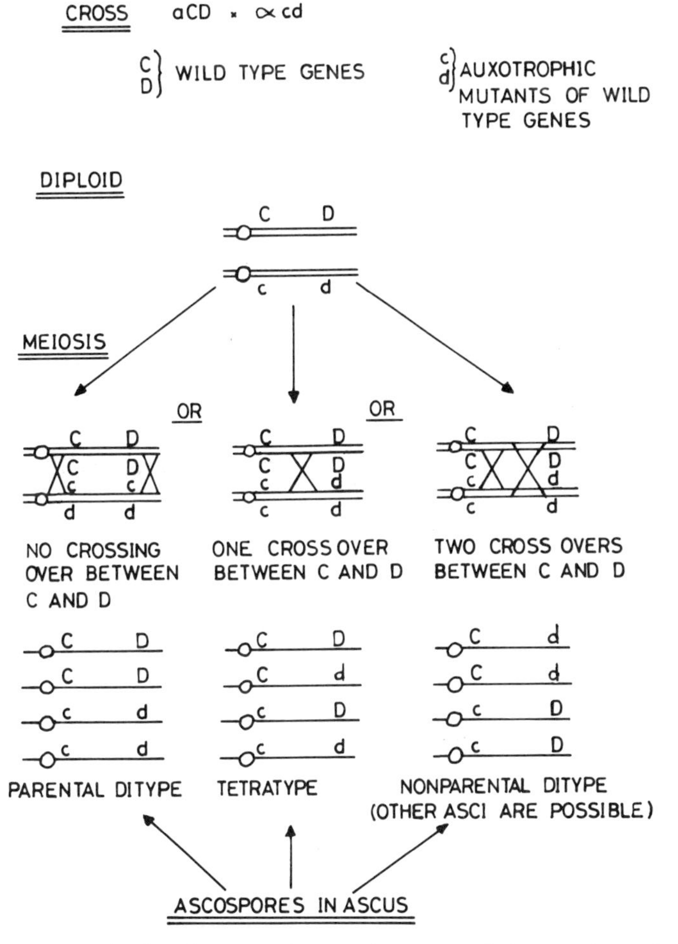

Fig. 7–2 Recombination of genes on the same chromosome during sexual reproduction in yeast.

tested for the different genetic markers used. The different results which can be obtained are shown in Fig. 7–2.

7.2.2 Mitotic recombination

Sometimes sectors appear in diploid colonies on a plate, or genetic changes occur in a population of cells growing in liquid culture. These may be the result of mutation but more often they are the result of a failure of the genome to replicate itself perfectly during mitosis. Several types of change come into this category and are known collectively as mitotic recombination.

7.3 Genetics of mitochondria

The development of mitochondrial genetics dates from the isolation by Ephrussi and Slonimski in 1953 of *petite* mutations. These strains gave smaller colonies when grown on a carbon source which can be metabolized anaerobically, and are unable to grow on medium containing glycerol or lactate as the sole carbon source. These compounds can only be metabolized aerobically. *Petite* mutants were found to lack several mitochondrial enzymes and were only able to metabolize carbon substrates by the fermentative route (Chapter 3).

Genetic studies indicated that the *petite* phenotype was sometimes caused by mutations in nuclear genes but could also be caused by mutation which did not segregate like nuclear genes during sexual reproduction, but exhibited cytoplasmic inheritance. These observations were made before the presence of DNA in mitochondria had been established. Subsequent work showed that the cytoplasmic *petite* mutations could be attributed to mutations in the mitochondrial DNA.

Since there are between twenty and thirty mitochondria in each yeast cell it is possible for a situation to arise in which a mixture of normal mitochondria, and mitochondria carrying a *petite* mutation in their DNA exist in the same cell. When this occurs, both mitochondria replicate at their own rate. If the *petite* mitochondria replicate at the same rate as the normal mitochondria, they are said to be neutral. If however the *petite* mitochondria divide faster, they slowly replace the wild type mitochondria. Such mutations are referred to as suppressive *petites* since they suppress the expression of the normal wild type mitochondria.

Other mitochondrial mutations have now been found. These include mutations which affect specific enzymes of the respiratory apparatus and also a series of antibiotic resistance mutations. Using these resistance markers it has been possible to demonstrate that recombination can occur between the DNA of different mitochondria.

7.4 Spheroplast fusion

Traditionally, genetic analysis could only be carried out in yeast strains which were able to complete the sexual cycle. Unfortunately many brewing strains are polyploid and do not readily undergo sexual reproduction. The genotype of such strains can now be manipulated using the technique of spheroplast fusion. In this process, two strains, each of which contain desirable characteristics are converted into spheroplasts, that is their cell wall is degraded and removed using enzymes. Since spheroplasts are very fragile and would burst in normal media, they are prepared in a medium containing 1.2 M sorbitol which has a similar osmotic strength to the cell itself. The spheroplasts from the two strains are then induced to fuse in a medium containing polyethyleneglycol and calcium chloride. The cells produced can then be stimulated to regenerate a new cell wall by suspending in a 3% agar medium. In this way, 'diploid' cells containing both parental sets of chromosomes are obtained. If appropriate mating type genes are present, these cells can then be induced to sporulate and produce 'haploid' ascospores in which the characteristics of the two strains have been reassorted.

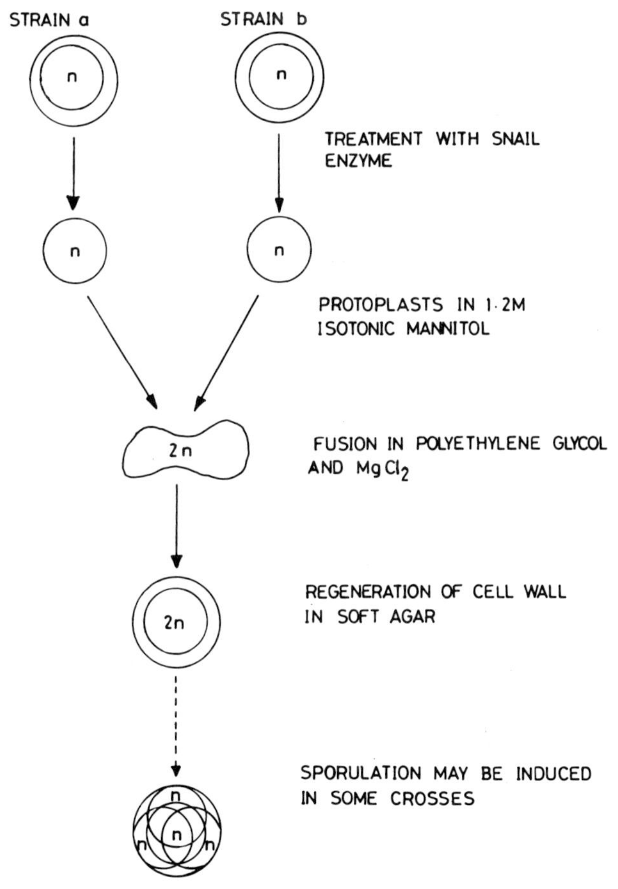

Fig. 7–3 Sequence of events used to achieve crossing in non fertile strains using spheroplast fusion.

Since this is a new technique, insufficient analysis has been carried out to confirm the precise mechanisms which lead to this reassortment. This technique offers many opportunities to study the genetics of strains of the same species, or even different species which cannot normally be induced to undergo normal plasmogamy and sexual reproduction (Fig. 7–3).

7.5 Transformation and gene transfer

Transformation is the name given to the process by which DNA can be extracted from one organism, the donor, and introduced into a second organism, the recipient, in which at least part of the new DNA becomes incorporated into the genetic material of the recipient. The process has been known for many years and intensively studied in bacteria. However, only in the past few years has convincing evidence that transformation can occur in yeast become available.

The process of transformation can be divided into three phases. The isolation of DNA from the donor strain, the transfer of the DNA into the recipient strain and finally the incorporation and expression of the new DNA in the recipient strain. It is possible to carry out transformation using a simple DNA extract which contains DNA for all potential donor genes; however more precise results, and a higher frequency of transformation are obtained if the DNA for the specific gene required can be used in a more purified form. This can be achieved by incorporating the DNA segment containing the required gene into circular nonchromosomal DNA molecules called plasmids which have been found in many bacteria and recently in yeast. The only plasmid DNA characterized in *Saccharomyces cerevisiae* is referred to as 2 μm circles. It is not appropriate in this short chapter to describe the techniques used to incorporate specific DNA segments into plasmids. It is sufficient to indicate that it is possible to produce a wide variety of plasmids containing many different genes for use in transformation experiments. At the time of writing, transformation has been carried out in yeast using a crude DNA extract, using bacterial plasmids and using plasmids derived from 2 μm circles of yeast.

Transfer of the DNA into the recipient strain is achieved by adding the DNA to a preparation of spheroplasts suspended in an isotonic solution containing polyethylene glycol. DNA molecules pass into the spheroplasts before cell wall regeneration is initiated.

Genes transferred into yeast cells by tranformation may become incorporated into nuclear chromosomes or remain in plasmids which replicate independently of the chromosomes and exhibit the characteristics of cytoplasmic inheritance. The procedure used to transform a cell which could not ferment maltotriose into one that could, is shown in Fig. 7–4.

7.6 Virus-like particles in yeast

The observation by Bevan and co-workers that some strains of yeast excreted a substance which killed other strains of yeast led to the discovery that yeasts,

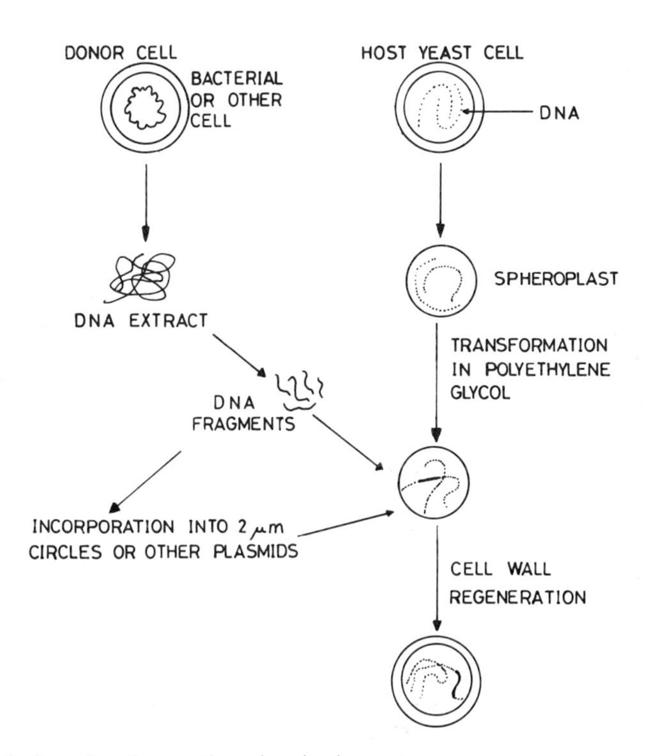

Fig. 7–4 Procedure for genetic engineering in yeast.

and in fact many other fungi, contain virus-like particles (VLPs). VLPs have been found to be present in all the strains which produce this 'killer' substance and many other strains which do not. VLPs consist of hollow spheres composed of protein subunits and containing one or two molecules of double-stranded RNA. The presence of RNA, which appears to act as the genetic material of the VLPs, is surprizing since double-stranded RNA molecules are unusual in biological systems and are generally restricted to a few plant and animal viruses. These VLPs resemble these viruses in many respects; however it is probably not correct to refer to them as viruses since it has not been demonstrated that they can be transferred from cell to cell by an infection process. Many genetic studies have been carried out on the killer system which may provide a model for the genetic control of double-stranded RNA viruses in plant and animal cells.

8 Yeasts in Industry

The origins and development of the use of yeast by man have been referred to previously (Chapter 1). At the present time, brewing and distilling industries exist legally or illegally in most countries in the world and constitute a major world industry. The value of alcoholic beverages produced has been assessed at £24×10^9 per year. Man indeed takes his pleasures very seriously!

The main property of yeast which has made it a valuable organism to man is its ability to convert high concentrations of sugars into alcohol with a high efficiency, giving rise to a product which has proved to be very palatable to man. The physiology of this process has been discussed in Chapter 3 and will not be covered again here; however, it is important to emphasize that the products of alcoholic fermentation are not just ethanol, but also include carbon dioxide and yeast biomass. Organoleptic compounds, produced by yeast, which add flavour to alcoholic beverages can also be viewed as important products of the fermentation.

A very large variety of alcoholic beverages are produced in different parts of the world. The majority are based on the fermentation of sugars by yeast and the differences arise from the source of the sugars fermented and whether the product is distilled or not. Some well known examples are shown in Table 2. Alcohol fermentations may achieve a final alcohol concentration of around 15% as in certain wines from Bordeaux. At these levels, ethanol is toxic to the yeast itself so if higher concentrations are required, the alcohol must be concentrated by distillation. However, most wines and beers have an alcohol concentration of 10% or less.

The fermentation of sugars produces almost as much carbon dioxide as ethanol. In the baking industry, it is the carbon dioxide

$$C_6H_{12}O_6 \rightarrow 2C_2H_5OH + 2CO_2$$

$$180\ g \rightarrow 2 \times 46\ g + 2 \times 44\ g$$

produced by the yeast which is the important product. The rising of the dough is caused by the production of carbon dioxide by yeast which has been trapped in the dough during the kneading process.

It is essential that the yeast is distributed evenly throughout the dough; if a bread with an even texture is to be obtained. The yeast also contributes to the flavour of the bread; however, this aspect is not normally emphasized and with modern active strains of bakers yeast, the level of yeast required is so low that it is rare to find bread with a yeasty flavour. Although carbon dioxide is a by-product of the distillation industry, it is collected in some large distilleries, compressed and sold as cylinders of liquid carbon dioxide. One market for bottled carbon dioxide is the beverage industry, which uses carbon dioxide to

Table 2

Beverage	Carbohydrate source	Type of carbohydrate	Source of amylase enzymes	Other source of flavours	Product distilled
Beer	Barley	Starch	Barley amylase	Hops	–
Whisky	Barley	Starch	Barley amylase	Peat, barrel	+
Vodka	Maize Potatoes	Starch Starch	Barley amylase and other		+
Gin	Maize Potatoes	Starch Starch	microbial enzymes	Juniper	+
Saké	Rice	Starch	*Aspergillus oryzae* amylase		+
Rum	Molasses	Sucrose	–		+
Wine	Grapes	Glucose Fructose	– –		–
Brandy	Grapes	Glucose Fructose	–		+
Cider	Apples	Glucose Fructose Sucrose	–		–
Calvados	Apples	Glucose Fructose Sucrose	–		+

produce fizzy drinks. This represents another example of carbon dioxide being an economically important product of the yeast fermentation.

During each fermentation, at least three times as much yeast is produced as was used to inoculate the fermentation. This excess of yeast represents another by-product which would be a waste if no use could be found for it. Traditionally, excess yeast from the brewing and distilling industries was used as bakers yeast. Distillers yeast was preferred since it lacked the hop flavours present in unwashed brewers yeast. This practice may still continue in many countries; however in most developed countries, special yeasts are grown up for use in the baking industry, and other outlets for brewers yeast have had to be found. One important outlet for this excess yeast is the production of hydrolysates and autolysates which are used as flavouring agents. Other 'spent' yeast is used in the preparation of animal feeds. Much of the yeast produced in the distilling industry is destroyed during the distillation stage and leaves the factory in the form of a thick brown effluent known as pot ale. This has been used for the production of animal feeds and, in the dried form known as distillers solubles as a nutrient source for other industrial fermentations.

The growth of yeast in anaerobic conditions leads to a high yield of ethanol, but a poor yield of yeast cells per unit substrate consumed. This method of growth is not suitable for any process in which a high yield of yeast cells is required such as the production of bakers yeast or the production of yeast biomass for use as an animal feed. The highest yields of yeast are obtained when the yeast is grown in highly aerobic conditions on a medium which contains a low concentration of sugars. Although industrial alcohol is now produced from oil, it has in the past been produced by fermentation. At present, only alcohol produced for human consumption is required to be produced by fermentation. In addition to alcoholic beverages, this includes some for pharmaceutical use and alcohol used as the starting material for vinegar production. A second chemical feedstock, glycerol, can also be produced by a yeast fermentation and during World War I, glycerol for use in the production of the explosive nitroglycerine, was produced by this route in Germany. At the present time, yeast fermentations are not an important source of organic solvents, however, the increases in oil prices recorded in the 1970s together with the anticipated world shortage of oil, have led to renewed interest in the utilization of renewable substrates such as sugar and starch for the production of ethanol by fermentation. In particular, the use of alcohol:petrol mixtures as a fuel for motor vehicles, is in operation in several countries and is being encouraged since alcohol can replace the environmentally dubious lead antiknock compounds at present used in many fuels.

8.1 Role of yeast in the production of alcoholic beverages

The production of most alcoholic beverages involves the fermentation of some carbohydrate source using strains of yeast which can be classified broadly as *Saccharomyces cerevisiae*, e.g. beer, cider, whisky, gin and saki production. However, many traditional wine fermentations use a mixture of yeasts and rum is frequently fermented using the fission yeast *Schizosaccharomyces*. The yeast makes a two-fold contribution to the production of an alcoholic beverage. It is responsible for the production of the ethanol in the beverage and also for the production of a variety of compounds which contribute to the flavour of the beverage. These are referred to as organoleptic compounds. The numbers of organoleptic compounds isolated from alcoholic beverages can be counted in hundreds. Many of these are present in very small quantities and are difficult to identify and quantify. It is even more difficult to assess the contribution of these different compounds to the final product since different compounds have different odour thresholds; that is the concentration at which the individual compounds can be detected by smell differs. Not only is the odour threshold unique to each compound but it is also influenced by the presence of other organoleptic compounds in the beverage.

8.2 Types of organoleptic compounds

Organoleptic compounds can be divided into at least five chemical groups. Higher alcohols, acids, esters, aldehydes and ketones and sulphur compounds,

(Fig. 8–1). A tracing from a gas liquid chromatograph analysis of an alcoholic beverage is shown in Fig. 8–2.

Since this book is concerned what yeast, emphasis has been placed on the organoleptic compounds produced by yeast. However, the final flavour of an

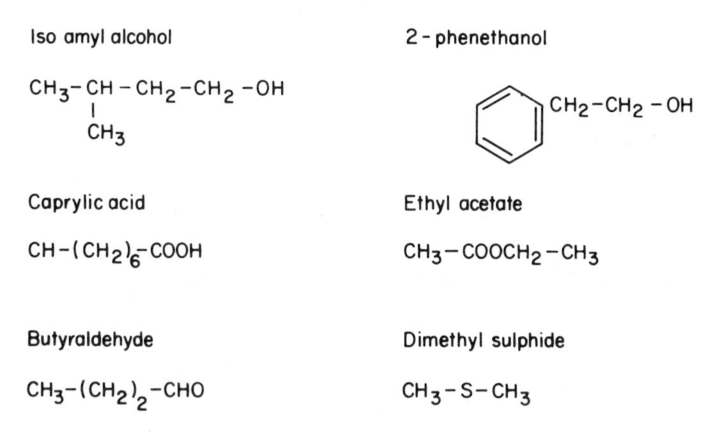

Fig. 8–1 Selected organoleptic compounds produced by yeast.

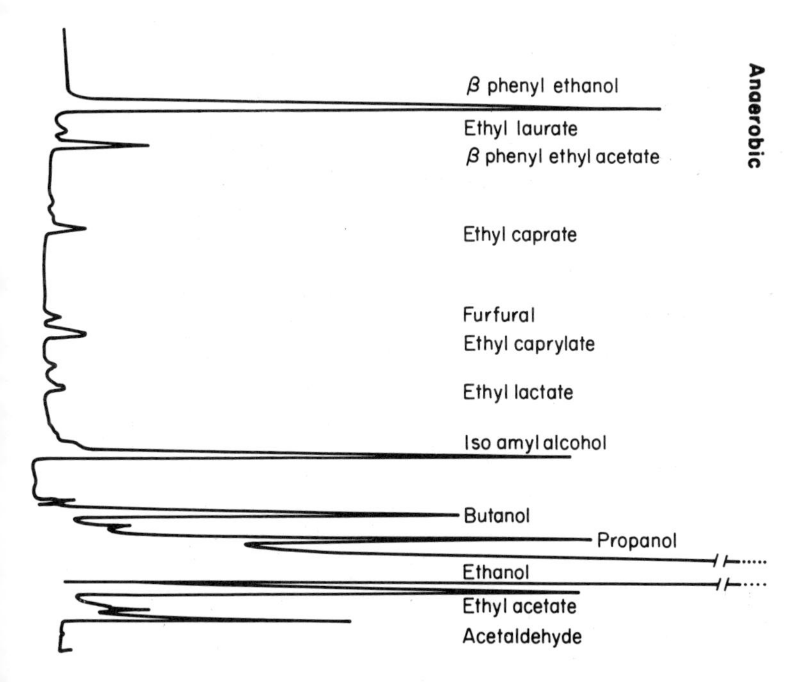

Fig. 8–2 Gas liquid chromatogram of organoleptic compounds in a distilled spirit.

alcoholic beverage is dependent upon a variety of factors. In wines and cider, the quality of the grapes and apples is of critical importance. In beer, the addition of hops to the fermentation makes a major contribution to the flavour and in whisky the treatment of the barley with peat smoke gives the final product a characteristic peaty aroma. In the production of gin, the distilled spirit is almost pure ethanol, and the flavour is provided by the addition of plant material or botanicals such as juniper berries and coriander to the spirit.

8.3 Fermentation processes

The production of alcoholic beverages can be divided into three or four stages. A flow diagram of the stages in the brewing of beer is shown in Fig. 8–3.

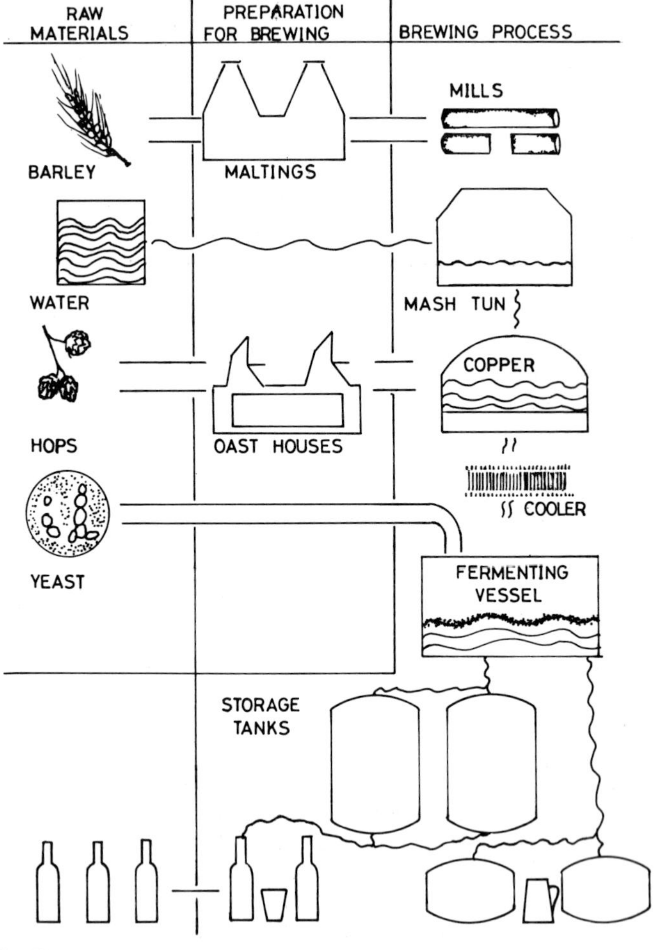

Fig. 8–3 Flow diagram of a typical brewing process.

8.3.1 *Preparation of raw materials*

Since the sugars present in grapes are readily assimilated by yeast, little preparation is required other than crushing the grapes to release the juice. In the production of red wine, the skins and seeds are left in contact with the grape juice throughout the fermentation, whereas in the production of white wines they are removed after the crushing stage. In contrast, the main carbohydrate of barley, starch, cannot be assimilated by yeast so an important part of the technology of brewing is the controlled germination of the barley to produce amylase enzymes. This process is known as malting and is usually carried out by specialist maltsters. The malted barley is then dried by kilning before being transferred to the brewery. During this drying process, coloured compounds may develop by the caramelization of the sugars which are important in giving the correct colour to beer. In the brewery, the malt is ground up and mixed with water in the mashing process. During this process the amylase enzymes break down most of the starch to metabolisable sugars such as maltose. Some of the starch, however, the dextrin fraction, is not broken down and is not metabolizable by yeast so it remains intact throughout the fermentation phase and gives body to the beer. The extract prepared during the mashing process, called wort, is filtered off from the debris of the barley seeds and used as the fermentation medium for beer production.

8.3.2 *The fermentation stage*

The fermentation phase in the production of most alcoholic beverages is normally a single stage batch process which can be carried out in simple wooden vats or in large stainless steel tanks. Wort for beer production is inoculated with a single strain of *Saccharomyces cerevisiae* which has been selected by the brewer to produce a beverage of the correct strength and flavour. In contrast, traditional wine fermentations are carried out by the natural flora from the skins of the grapes so a wide variety of species contribute to the fermentation, e.g. *Brettanomyces* sp., *Candida* sp., *Debaryomyces*, and *Hansenula*, as we as *Saccharomyces*. After inoculation, the fermentation proceeds with a rapid growth phase which is associated with a rapid period of ethanol formation and sugar consumption. After this period, the rate of alcohol formation slows down. The fermentation period may last from a few days in beer production to several weeks in wine production.

8.3.3 *Maturation and distillation*

Beers, lagers and some wines are matured for a few weeks, whereas many wines and the distillates used for the production of good whisky and brandy may be matured for several years. In beer production, the maturation phase is important for the removal of certain undesirable volatile components, and for the precipitation of polyphenolic compounds which can give rise to turbidity in beer. In the maturation of wine, malic acid is removed from the wine by the growth of bacteria in the so-called malo-lactic fermentation. This and other biochemical changes lead to an improved flavour in the wine during maturation.

Different distillation procedures have been used to produce different spirits and the use of the correct apparatus and procedure is essential to the formation of the best quality spirits. During the period of maturation of spirits such as whisky and brandy some products are extracted from the wood of the casks which are used to store the spirit during maturation. These can play an important role in the development of the flavour.

8.4 Gasohol production

In theory, ethanol can be produced from any carbohydrate source which can be converted into a form which can be metabolized by yeast. In practice, economics dictates that the substrate must be easily converted into metabolizable form. In a major programme in Brazil two sources of carbohydrate have been used; sucrose in the form of an extract from sugar cane, and starch from cassava. Since starch cannot be fermented directly by *Saccharomyces cerevisiae*, it is hydrolysed to glucose and maltose using industrial amylases produced from other micro-organisms by fermentation and not barley amylases. Starch from other sources such as maize has been used in other countries for bulk ethanol production. One genuine concern at the use of sucrose and starch products for ethanol production is that these are substrates which are also required for food for man and animals so extensive production of alcohol could result in food shortages in developing countries. Cellulose is the most abundant form of carbohydrate and is not used as a food product. This would be an ideal substrate for ethanol production if it could be hydrolysed to glucose at a low price. Chemical and enzymic hydrolysis have been tested and found too expensive at present, so current research is directed at obtaining cheaper processes for the hydrolysis of cellulose. At present, the production of industrial alcohol from cellulose is not viable and the economics of its production from sucrose and starch are dubious. However, some countries which have to import oil believe that the production of alcohol by fermentation as a substitute for gasoline is advantageous since it reduces the expenditure of hard currency on the importation of oil.

The assessment of the value of a process for producing an energy product such as ethanol is very complex. Although 1 kg of ethanol has a gross calorific value of 29.7 MJ, only a part of that is a net energy gain, since some energy has been put into the system during the growth of the crop, the transport of the crop to the distillery and the distillation process itself. In spite of these complexities, it is anticipated that such processes will make a contribution to the energy needs of several countries in the future.

8.5 Bakers yeast and biomass production

Although the primary objective of bakers yeast production is to obtain a yeast which produces carbon dioxide at a high rate in dough, it can be considered as a specialized process of biomass production. Since bakers yeast is added to flour at a concentration of 1% on a weight basis, it represents an important source of microbial biomass in human diets. A value of 2 g protein per person per week has been estimated for yeast protein in the western European diet. The amino acid composition and vitamin content of bakers yeast is shown in Table 3. Although yeast has been found to be an acceptable food additive for many years, it is inferior in nutritional quality to animal proteins. However, the traditional use of yeast in baking has resulted in it being more readily acceptable than other micro-organisms as a food source. Although it appears to be largely free of toxic side effects, it does contain a relatively high level of nucleic acid. Since high levels of nucleic acid can lead to high levels of uric acid in man, which causes gout, an upper limit of 30 g dried yeast per day has been recommended.

The production of bakers yeast involves the solution of several dilemmas. Although the product is required to work in an anaerobic environment, it must be produced in a highly aerobic environment so that good yields can be obtained. The yeast produced must have a high activity in the dough, but also a good storage capacity and in the case of dried yeast, a good drying characteristic. Unfortunately the strains exhibiting the highest fermentation rates tend to have poor storage and drying properties, so a compromise set of growth conditions has to be used to produce a bakers yeast which has a good activity and good stability. The yeast is grown up in a series of large vessels which are well mixed and aerated and fed with a nutrient solution containing sugars, salts and vitamins. This is usually based on molasses. The nutrient mixture is fed either continuously or at short intervals throughout the fermentation, rather than being added in its entirety at the beginning of the fermentation. If too much sugar is added at any one time, the sugar level rises, yeast growth becomes fermentative and the yield is reduced (see Chapter 3). When growth has been completed, the yeast is concentrated by centrifugation, then filtered to produce a cake which can be pressed into blocks of compressed yeast. Dried yeast is prepared by drying yeast cake in drum driers or more recently in fluidized bed driers.

Other yeasts have been used in a variety of processes to produce biomass from waste carbohydrates. The use of *Candida utilis* to produce biomass from sulphite liquor, a waste product from the paper industry, was favoured because it can utilize xylose, a sugar present in sulphite liquor which *Saccharomyces* strains cannot use. *Endomycopsis fibuliger* has been used to produce biomass from effluents containing starch, and the yeast *Kluyveromyces lactis* which can metabolize lactose, for the production of biomass from whey. An interesting development in the last decade has been the growth of yeasts on C_{10} – C_{20} liquid alkanes which can be obtained from mineral oil and on methanol produced by the chemical oxidation of natural gas.

8.6 Yeast derived products

8.6.1 Autolysates and hydrolysates

Yeast hydrolysates and autolysates have the ability either to enhance or to impart a meaty flavour to food products, so they are widely used in the food industry in the preparation of soups and sauces, and in the flavouring of snacks such as potato crisps. Hydrolysates are produced by heating the yeast cells to 100°C together with HCl, until most of the protein has been hydrolysed to amino acids. The preparation is then neutralized with NaOH, filtered and concentrated to a thick paste. The final product contains a high level of salt, produced by the neutralization of the acid.

Autolysis differs from hydrolysis, in that the breakdown of the cell constituents such as proteins and nucleic acids is achieved by the action of enzymes produced by the yeast cell itself. This process can occur naturally, but is facilitated by heating to 50°C and the addition of salt. Autolysis normally proceeds for one day, until at least half the protein has been broken down to amino acids. The product is then filtered and concentrated to a thick paste. It is probable that the meaty flavour which is characteristic of yeast autolysates, can be attributed to the amino acids and small peptides produced by the action of proteases during the autolysis, although nucleotides such as 5'inosine monophosphate, and 5'guanosine monophosphate could also play an important role in flavour enhancement.

8.6.2 Enzymes

The enzyme invertase has found a wide use in the confectionery industry in the production of soft-centred chocolates. The flavoured centre is prepared as a solid shape of crystalline sucrose containing flavouring and a small amount of

Table 3 Nutritional value of yeast.

Essential amino acids	(g 100g dry wt^{-1})	*Vitamins*	(µg g dry wt^{-1})
Lysine	8.2	Thiamine HCl	165
Valine	5.5	Riboflavin	100
Leucine	7.9	Niacin	585
Isoleucine	5.5	Pyridoxine HCl	20
Threonine	4.8	Folic acid	13
Methionine	2.5	Calcium pantothenate	100
Phenylalanine	4.5	Biotin	0.6
Tryptophan	1.2	Paraaminobenzoic acid	160
Cystine	1.6	Choline chloride	2710
Histidine	4.0	Inositol	3000
Tyrosine	5.0		
Arginine	5.0		

invertase. The solid shape is then coated with chocolate. During storage, the invertase acts on the sucrose, converting it to a mixture of fructose and glucose, thus turning the crystalline centre to a syrup.

High levels of invertase (sucrase) are present in yeast cells such as bakers yeast which have been grown on molasses. Molasses is a by-product of the sugar industry and is rich in sucrose. The enzyme, which occurs both in the cytoplasm of the cell and bound to the cell wall, is released by autolysing the cells, purified to a limited extent, and sold to the confectionery industry as a standard solution of known invertase activity. Although yeast contains many other enzymes, some of which are extracted and purified for use as laboratory enzymes, no other enzymes of *Saccharomyces cerevisiae* are produced on a commercial scale. The enzyme lactase, which converts lactose to glucose and galactose, has however been produced commercially from another yeast, *Kluyveromyces lactis*.

8.6.3 Vitamins

Yeasts are a rich source of water soluble vitamins (see Table 3). These are normally sold in the form of tablets, prepared from dried yeast.

8.7 Coda

The development of the yeast based industries referred to in this book have been associated with a series of major advances both in the technology of handling organisms on an industrial scale and in our understanding of the fundamental processes of living organisms. Technology has benefited from fundamental studies such as the isolation of pure cultures, the development of yeast genetics and the elucidation of the control of yeast metabolism. Many of the techniques which are now recognized as essential to the modern subject of biotechnology such as the development and storage of industrial strains of micro-organisms, the use of fed batch cultures, the use of continuous culture and tower fermenters, have found an early application in yeast industries.

At the present time, the traditional food and beverage related industries dominate yeast technology; however, there is good reason to believe that this may not be the case in the future. Even a small replacement of petrol with fermented alcohol could lead to a major industry. Even at the present time, large amounts of yeast biomass are being produced by the beer and wine industries and there is an urgent need to develop new processes for utilizing this valuable product.

Yeast offers many advantages as an organism for genetic engineering and there are already indications that it is becoming the preferred organism for the production of pharmaceutical products by genetic engineering. Some, e.g. interferon, have already been produced in the laboratory using yeast clones. Yeast technology has an impressive past but looks to have an even brighter future.

Appendix

Media for yeast studies

(i) Complete medium for growth:	M.Y.G.P.
Malt extract	0.3% w/v
Yeast extract	0.3% w/v
D glucose	1.0% w/v
Mycological peptone	0.5% w/v
	in distilled water*
(ii) Sporulation medium	
Sodium acetate	0.7%
	in distilled water*

*To prepare solid medium add 2% agar.

Further Reading

CARTER, B.L.A. (1978). The yeast nucleus. In *Advances in Microbial Physiology*. Eds. A.H. Rose and J.G. Harrison, **17**, 243–302. Academic Press.

HARTWELL, L.H. (1974). *Saccharomyces cerevisiae*: cell cycle. *Bacteriological Reviews*, **38**, 164–98.

PETES, T.D. (1980). Molecular genetics of yeast. *Annual Review Biochemistry*, **49**, 845–76.

PRESCOTT, D.M. (1975). Methods in cell biology. Vols 11 and 12. *The Yeast Cell*. Academic Press.

REED, G. and PEPPLER, J.J. (1973). *Yeast Technology*. AVI Publishing Co., Westport, Connecticut, USA.

ROSE, A.H. (1977). Economic microbiology. *Vol. 1 Alcoholic Beverages*. Academic Press.

ROSE, A.H. and HARRISON, J.S. *The Yeasts*. Academic Press

Vol. 1 Biology (1969)

Vol. 2 Physiology and Biochemistry (1971)

Vol. 3 Yeast Technology (1970).

SUOMALAINEN, H., NURMINEN, J. and OURA, E. (1973). Aspects of cytology and metabolism of yeast. *Progress in Industrial Microbiology*, **12**, 109–67.

Index

Ascospores 2, 3, 32, 34, 37, 43
Ascus 37, 39
Autolysates 55–56
Baker's yeast 54–55
Baking 1, 48
Beer 1, 3, 48, 49, 50–53
Biomass 48
Birth scar 5, 8, 26
Bread 1
Brewing 1, 3
Bud scar 5, 8, 26, 27
Budding 3, 4, 5, 23–24, 26–27, 30, 32, 33, 34
Catabolite repression 11, 15, 16
cdc mutants 29, 30, 39
Cell cycle 23–24, 27–31, 34, 35, 36, 39, 40
Cell division 5
Cell membrane 6
Cell wall 4, 5, 6, 7, 17, 23–24, 34
Cell wall synthesis 25–26
Chitin 5, 25, 27
Chromatic bodies 7
Chromosomes 17–18, 36
Cider 3, 49, 50–53
Classification 2, 3
Differentiation 39
Distilled spirits 49, 51
Distilling 3
DNA 17–19, 22, 28, 35, 39, 46
DNA division cycle 29
DNA replication 18–19, 24, 40
Effluent 49
Enzymes 56
Essential amino acids 56
Fermentation 2, 11, 13–15, 48–49, 52
Flocculence 5
Fusel alcohols 2
Gasohol 54
Genetic engineering 57
Genetics 41–47
Genetic markers, selection of, 41
Glucan 5, 6, 25
Glycerol 50
Growth 10
Growth cycle 30
Hydrolysates 55–56
Industrial alcohol 50, 54
Invertase 25
Killer system 47
Life cycle 2, 32
Mannan 5, 6, 25
Mating 41
Mating types 32, 33
Meiosis 35, 37, 42
Mitochondria 8, 21–22, 34
Mitochondria, genetics of 44
Mitochondrial DNA 44
Mitotic recombination 44
Nucleus 7, 24
Organoleptic compounds 2, 48, 50–51
Pasteur effect 11, 14, 16
petite mutations 46
Plasmid, 2 μm 17, 46
Plasmogamy 33, 40
Protein biosynthesis 20–21
Recombination 42–44
Respiration 11, 12–13, 15
RNA 19–20, 28, 38, 47
RNA biosynthesis 20
Sexual reproduction 2
Spheroplast fusion 44–46
Spindle plaque 7, 23–24, 30, 34, 37
Spore germination 40
Sporulation 32, 34–41
Synaptonemal complexes 39
Transformation 46
Vacuole 9, 36
Virus-like particles 46
Vitamins 56–57
Whisky 1, 49, 50–53
Wine 1, 3, 48, 49–53

QK 617.5 .B48 1982